Lecture Notes in Computer Science 8455

Commenced Publication in 1973
Founding and Former Series Editors:
Gerhard Goos, Juris Hartmanis, and Jan van Leeuwen

Elisabeth Métais   Mathieu Roche
Maguelonne Teisseire (Eds.)

Natural Language Processing and Information Systems

19th International Conference on Applications
of Natural Language to Information Systems, NLDB 2014
Montpellier, France, June 18-20, 2014
Proceedings

 Springer

Volume Editors

Elisabeth Métais
Conservatoire National des Arts et Métiers
Department of Computer Science
2 rue Conté
75003 Paris, France
E-mail: metais@cnam.fr

Mathieu Roche
Cirad, TETIS
500 rue J.F. Breton
34093 Montpellier Cedex 5, France
E-mail: mathieu.roche@cirad.fr

Maguelonne Teisseire
Irstea, TETIS
500 rue J.F. Breton
34093 Montpellier Cedex 5, France
E-mail: maguelonne.teisseire@irstea.fr

ISSN 0302-9743   e-ISSN 1611-3349
ISBN 978-3-319-07982-0   e-ISBN 978-3-319-07983-7
DOI 10.1007/978-3-319-07983-7
Springer Cham Heidelberg New York Dordrecht London

Library of Congress Control Number: 2014940337

LNCS Sublibrary: SL 3 – Information Systems and Application, incl. Internet/Web
and HCI

Typesetting: Camera-ready by author, data conversion by Scientific Publishing Services, Chennai, India

Printed on acid-free paper

Springer is part of Springer Science+Business Media (www.springer.com)

Preface

This volume of *Lecture Notes in Computer Science* (LNCS) contains the papers selected for presentation at the 19th International Conference on Application of Natural Language to Information Systems, held in Montpellier, France (TETIS laboratory and CIRAD) during June 18–20, 2014 (NLDB2014). Since its foundation in 1995, the NLDB conference has attracted state-of-the-art works and followed closely the developments of the application of natural language to databases and information systems in the wider meaning of the term.

The current conference proceedings reflect the development in the field and encompass areas such as syntactic, lexical, and semantic analysis; information extraction; information retrieval; Social Networks, Sentiment Analysis, and other Natural Language Analysis. This year, at the initiative of Dr. Eric Kergosien, a demo session was proposed and the associated papers focused on "Semantic Information, Visualization and Information Retrieval." NLDB is now an established conference and attracts researchers and practitioners from all over the world. Indeed, this year NLDB had submissions from 26 countries. Moreover, papers dealt with a large number of natural languages that include Vietnamese, Arabic, Tunisian, Italian, Spanish, and German.

The contributed papers were selected from 73 papers (61 regular papers and 12 demo papers) and each paper was reviewed by at least three reviewers. The conference and Program Committee co-chairs had a final consultation meeting to look at all the reviews and made the final decisions. The accepted papers were divided into four categories: 12 papers as long/regular papers, 8 short papers, 12 posters, and 7 demos.

This year, we had the opportunity to welcome two invited presentations:

- Dr. Gabriella Pasi (University of Milano Bicocca, Italy)
  Personal Ontologies
- Pr. Sophia Ananiadou (University of Manchester, UK)
  Applications of Biomedical Text Mining

We would like to thank all who submitted papers for reviews and for publication in the proceedings and all the reviewers for their time, effort, and for completing their assignments on time.

We are also grateful to Cirad, AgroParisTech, La Région Languedoc Roussillon for sponsoring the event.

April 2014

Elisabeth Métais
Mathieu Roche
Maguelonne Teisseire

Organization

Conference Chairs

Elisabeth Métais   Conservatoire National des Arts et Metiers, Paris, France

Program Committee Chairs

Mathieu Roche   TETIS, Cirad, France
Maguelonne Teisseire   TETIS, Irstea, France

Demo Committee Chairs

Mathieu Roche   TETIS, Cirad, France
Maguelonne Teisseire   TETIS, Irstea, France
Eric Kergosien   LIRMM, France

Program Committee

Ben Hamadou Abdelmajid	Sfax University, Tunisia
Jacky Akoka	CNAM, France
Frederic Andres	National Institute of Informatics, Japan
Imran Sarwar Bajwa	University of Birmingham, UK
Nicolas Béchet	IRISA, France
Johan Bos	Groningen University, The Netherlands
Goss Bouma	Groningen University, The Netherlands
Sandra Bringay	LIRMM, France
Philipp Cimiano	Universität Bielefeld, Germany
Isabelle Comyn-Wattiau	ESSEC, France
Atefeh Farzindar	NLP Technologies Inc., Canada
Vladimir Fomichov	Faculty of Business Informatics, State University, Russia
Prószéky Gábor MorphoLogic	Pázmány University, Hungary
Alexander Gelbukh	Mexican Academy of Science, Mexico
Yaakov Hacohen-Kerner	Jerusalem College of Technology, Israel
Dirk Heylen	University of Twente, The Netherlands
Helmut Horacek	Saarland University, Germany
Dino Ienco	TETIS, France
Diana Inkpen	University of Ottawa, Canada

Maguelonne Teisseire   Irstea, TETIS, France
Bernhard Thalheim   Kiel University, Germany
Michael Thelwall   University of Wolverhampton, UK
Krishnaprasad Thirunarayan   Wright State University, USA
Juan Carlos Trujillo   Universidad de Alicante, Spain
Christina Unger   Universität Bielefeld, Germany
Alfonso Ureña   University of Jaén, Spain
Sunil Vadera   University of Salford, UK
Panos Vassiliadis   University of Ioannina, Greece
Roland Wagner   Linz University, Austria
René Witte   Concordia University, Canada
Magdalena Wolska   Saarland University, Germany
Erqiang Zhou   University of Electronic Science and
                                Technology, China

The following colleagues also reviewed papers for the conference and are due our special thanks:

Armando Collazo   Perea Ortega
Hector Dávila Diaz   Solen Quiniou
Elnaz Davoodi   Bahar Sateli
Cedric Du Mouza   Wenbo Wang
Kalpa Gunaratna   Nicola Zeni
Ludovic Jean-Louis   Yenier Castañeda Alexandre
Rémy Kessler   Chávez López
Sarasi Lalithsena   Miguel A. García Cumbreras
Jess Peral   Eugenio Martínez-Cámara
José Manuel

Organisation Committee

Hugo Alatrista Salas   TETIS and LIRMM, France
Jérôme Azé   LIRMM, France
Soumia Lilia Berrahou   LIRMM and INRA, France
Flavien Bouillot   LIRMM, France
Sandra Bringay   LIRMM, France
Sophie Bruguiere Fortuno   Cirad, TETIS, France
Juliette Dibie   AgroParisTech, France
Annie Huguet   Cirad, TETIS, France
Dino Ienco   Irstea, TETIS, France
Eric Kergosien   TETIS and LIRMM, France
Juan Antonio Lossio   LIRMM, France
Elisabeth Métais   Conservatoire National des Arts et Métiers,
                                Paris
Pascal Poncelet   LIRMM, France

Invited Speakers

Personal Ontologies

Gabriella Pasi

Università degli Studi di Milano Bicocca
Department of Informatics, Systems and Communication
Viale Sarcha 336,20131 Milano, Italy
gabriella.pasi@unimib.it

Abstract. The problem of defining user profiles has been a research issue since a long time; user profiles are employed in a variety of applications, including Information Filtering and Information Retrieval. In particular, considering the Information Retrieval task, user profiles are functional to the definition of approaches to Personalized search, which is aimed at tailoring the search outcome to users. In this context the quality of a user profile is clearly related to the effectiveness of the proposed personalized search solutions. A user profile represents the user interests and preferences; these can be captured either explicitly or implicitly. User profiles may be formally represented as bags of words, as vectors of words or concepts, or still as conceptual taxonomies. More recent approaches are aimed at formally representing user profiles as ontologies, thus allowing a richer, more structured and more expressive representation of the knowledge about the user. This talk will address the issue of the automatic definition of personal ontologies, i.e. user-related ontologies. In particular, a method that applies a knowledge extraction process from the general purpose ontology YAGO will be described. Such a process is activated by a set of texts (or just a set of words) representatives of the user interests, and it is aimed to define a structured and semantically coherent representation of the user topical preferences. The issue of the evaluation of the generated representation will be discussed too.

Applications of Biomedical Text Mining

Sophia Ananiadou

School of Computer Science, National Centre for Text Mining,
University of Manchester

Abstract. Given the enormous amount of biomedical literature produced, text mining methods are needed to extract, categorise, and curate meaningful knowledge from it. Biomedical text mining has focussed on the automatic recognition, categorisation, and normalisation of entities and their variant forms, which facilitated entity-based searching of documents, a far more effective method than simple keyword-based search. The extraction of relationships between entities has been applied to the discovery of potentially unknown associations between different biomedical concepts. Recently, event recognition has become a major focus of biomedical text mining research. Community challenges such as BioNLP has been a major factor in the increasing sophistication of event extraction systems, both in terms of the complexity of the information extracted and the coverage of different biological subdomains.

Moving beyond the simple identification of pairs of interacting proteins in restricted domains, state-of-the art systems can recognise and categorise various types of events (positive/negative regulation, binding, etc.), and a range of different participants relating to the reaction. Furthermore, different textual and discourse contexts of events result in different interpretations, i.e., hypotheses, proven observations, negation, tentative analytical conclusions, etc. Systems able to recognise and capture these interpretations provide opportunities to develop more sophisticated applications. Biomedical text mining applications include more focussed and relevant search, pathway reconstruction, linking clinical with biomedical information for clinical systems, detecting potential contradictions or inconsistencies in information reported in different articles, etc. All these applications are facilitated by interoperable Web-based text mining platforms such as NaCTeM's Argo, which support the customisation of solutions through modularization and task specific parameterization of text mining workflows.

Table of Contents

Syntactic, Lexical and Semantic Analysis

Information Extraction

Information Retrieval

Social Networks, Sentiment Analysis, and other Natural Language Analysis

Demonstration Papers

Using Wiktionary to Build an Italian
Part-of-Speech Tagger

Tom De Smedt[1], Fabio Marfia[2], Matteo Matteucci[2], and Walter Daelemans[1]

[1] CLiPS Computational Linguistics Research Group, University of Antwerp,
Antwerp, Belgium
`tom@organisms.be`, `walter.daelemans@uantwerpen.be`
[2] DEIB Department of Electronics, Information and Bioeng., Politecnico di Milano,
Milan, Italy
{`marfia,matteucci`}`@elet.polimi.it`

Abstract. While there has been a lot of progress in Natural Language
Processing (NLP), many basic resources are still missing for many languages, including Italian, especially resources that are free for both research and commercial use. One of these basic resources is a Part-of-Speech
tagger, a first processing step in many NLP applications. We describe a
weakly-supervised, fast, free and reasonably accurate part-of-speech tagger for the Italian language, created by mining words and their part-of-
speech tags from Wiktionary. We have integrated the tagger in Pattern,
a freely available Python toolkit. We believe that our approach is general
enough to be applied to other languages as well.

Keywords: natural language processing, part-of-speech tagging, Italian, Python.

1 Introduction

A survey on Part-of-Speech (POS) taggers for the Italian language reveals only
a limited number of documented resources. We can cite an Italian version of
TreeTagger [1], an Italian model[1] for OpenNLP [2], TagPro [3], CORISTagger
[4], the WaCky corpus [5], Tanl POS tagger [6], and ensemble-based taggers
([7] and [8]). Of these, TreeTagger, WaCky and OpenNLP are freely available
tools. In this paper we present a new free POS tagger. We think it is a useful
addition that can help the community advance the state-of-the art of Italian
natural language processing tools. Furthermore, the described method for mining
Wiktionary could be useful to other researchers to construct POS taggers for
other languages.

The proposed POS tagger is part of Pattern. Pattern [9] is an open source
Python toolkit for data mining, natural language processing, machine learning
and network analysis, with a focus on user-friendliness (e.g., users with no expert
background). Pattern contains POS taggers with language models for English
[10], Spanish [11], German [12], French [13] and Dutch[2], trained using Brill's

---

[1] `https://github.com/aparo/opennlp-italian-models`
[2] `http://cosmion.net/jeroen/software/brill_pos/`

E. Métais, M. Roche, and M. Teisseire (Eds.): NLDB 2014, LNCS 8455, pp. 1–8, 2014.

tagging algorithm (except for French). More robust methods have been around for some time, e.g., memory-based learning [14], averaged perceptron [15], and maximum entropy [16], but Brill's algorithm is a good candidate for Pattern because it produces small data files with fast performance and reasonably good accuracy.

Starting from a (manually) POS-tagged corpus of text, Brill's algorithm produces a lexicon of known words and their most frequent part-of-speech tag (aka a tag dictionary), along with a set of morphological rules for unknown words and contextual rules that update word tags according to the word's role in the sentence, considering the surrounding words. To our knowledge the only freely available corpus for Italian is WaCky, but, being produced with TreeTagger, it does not allow commercial use. Since Pattern is free for commercial purposes, we have resorted to constructing a lexicon by mining Wiktionary instead of training it with Brill's algorithm on (for example) WaCky. Wiktionary's GNU Free Documentation License (GFDL) includes a clause for commercial redistribution. It entails that our tagger can be used commercially; but when the data is modified, the new data must again be "forever free" under GFDL.

The paper proceeds as follows. In Sect. 2 we present the different steps of our data mining approach for extracting Italian morphological and grammatical information from Wiktionary, along with the steps for obtaining statistical information about word frequency from Wikipedia and newspaper sources. In Sect. 3 we evaluate the performance of our tagger on the WaCKy corpus. Sect. 4 gives an overview of related research. Finally, in Sect. 5 we present some conclusions and future work.

2   Method

In summary, our method consists of mining Wiktionary for words and word part-of-speech tags to populate a lexicon of known words (Sect. 2.1 and 2.2), mining Wikipedia for word frequency (Sect. 2.3), inferring morphological rules from word suffixes for unknown words (Sect. 2.5), and annotating a set of contextual rules (Sect. 2.6). All the algorithms described are freely available and can be downloaded from our blog post[3].

2.1   Mining Wiktionary for Part-of-Speech Tags

Wiktionary is an online "collaborative project to produce a free-content multilingual dictionary"[4]. The Italian section of Wiktionary lists thousands of Italian words manually annotated with part-of-speech tags by Wiktionary contributors. Since Wiktionary's content is free we can parse the HTML of the web pages to automatically populate a lexicon.

---

[3] http://www.clips.ua.ac.be/pages/using-wiktionary-to-build-an-italian-part-of-speech-tagger

[4] http://www.wiktionary.org/

We mined the Italian section of Wiktionary[5], retrieving approximately a 100,000 words, each of them mapped to a set of possible part-of-speech tags. Wiktionary uses abbreviations for parts-of-speech, such as *n* for nouns, *v* for verbs, *adj* for adjectives or *n v* for words that can be either nouns or verbs. We mapped the abbreviations to the Penn Treebank II tagset [20], which is the default tagset for all taggers in Pattern. Since Penn Treebank tags are not equally well-suited to all languages (e.g., Romance languages), Pattern can also yield universal tags [21], automatically converted from the Penn Treebank tags. Some examples of lexicon entries are:

- *di* → IN (preposition or subordinating conjunction)
- *la* → DT, PRP, NN (determiner, pronoun, noun)

Diacritics are taken into account, i.e., *di* is different from *dì*. The Italian section of Wiktionary does not contain punctuation marks however, so we added common punctuation marks (?!.::,()[]+-*\) manually.

2.2   Mining Wiktionary for Word Inflections

In many languages, words inflect according to tense, mood, person, gender, number, and so on. In Italian, the plural form of the noun *affetto* (*affection*) is *affetti*, while the plural feminine form of the adjective *affetto* (*affected*) is *affette*. Unfortunately, many of the inflected word forms do not occur in the main index of the Italian section of Wiktionary. We employed a HTML crawler that follows the hyperlink for each word and retrieves all the inflected word forms from the linked detail page (e.g., conjugated verbs and plural adjectives according to the gender). The inflected word forms then inherit the part-of-speech tags from the base word form. We used simple regular expressions to disambiguate the set of possible part-of-speech tags, e.g., if the detail page mentions *affette* (plural adjective), we did not inherit the *n* tag of the word *affetto* for this particular word form.

Adding word inflections increases the size of the lexicon to about 160,000 words.

2.3   Mining Wikipedia for Texts

We wanted to reduce the file size, without impairing accuracy, by removing the "less important" words. We assessed a word's importance by counting how many times it occurs in popular texts. A large portion of omitted, low-frequency words can be tagged using morphological suffix rules (see Sect. 2.5).

We used Pattern to retrieve articles from the Italian Wikipedia with a spreading activation technique [22] that starts at the *Italia* article (i.e., one of the top[6] articles), then retrieves articles that link to *Italia*, and so on, until we reached

---

[5] http://en.wiktionary.org/wiki/Index:Italian
[6] http://stats.grok.se/it/top

1M words (=600 articles). We boosted the corpus with 1,500 recent news articles and news updates, for another 1M words. This biases our tagger to modern Italian language use.

We split sentences and words in the corpus, counted word occurrences, and ended up with about 115,000 unique words mapped to their frequency. For example, *di* occurs 70,000 times, *la* occurs 30,000 times and *indecifrabilmente* (*indecipherable*) just once. It follows that *indecifrabilmente* is an infrequent word that we can remove from our lexicon and replace with a morphological rule, without impairing accuracy:

-*mente* → RB (adverb).

Morphological rules are discussed further in Sect. 2.5.

2.4   Preprocessing a CSV File

We stored all of our data in a Comma Separated Values file (CSV). Each row contains a word form, the possible Penn Treebank part-of-speech tags, and the word count from Sect. 2.3 (Table 1).

Table 1. Top five most frequent words

Word form	Parts of speech	Word Count
di	IN	71,655
e	CC	44,934
il	DT	32,216
la	DT, PRP, NN	29,378
che	PRP, JJ, CC	26,998

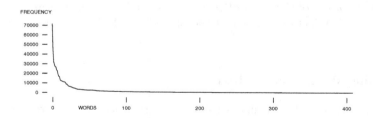

Fig. 1. Frequency distribution along words

The frequency distribution along words is shown in Fig. 1. It approximates Zipf's law: the most frequent word *di* appears nearly twice as much as the second most frequent word *e*, and so on. The top 10% most frequent covers 90% of popular Italian language use (according to our corpus). This implies that we can remove part of "Zipf's long tail" (e.g., words that occur only once). If we have a lexicon that covers the top 10% and tag all unknown words as NN (the most frequent tag), we theoretically obtain a tagger that is about 90% accurate. We were able to improve the baseline by about 3% by determining appropriate morphological and contextual rules.

2.5   Morphological Rules Based on Word Suffixes

By default, the tagger will tag unknown words as NN. We can improve the tags of unknown words using morphological rules. One way to predict tags is to look at word suffixes. For example, English adverbs usually end in *-ly*. In Italian they end in *-mente*. In Table 2 we show a sample of frequent suffixes in our data, together with their frequency and the respective tag distribution.

We then automatically constructed a set of 180 suffix rules based on high coverage and high precision, with some manual supervision. For example:

$$-mente \rightarrow RB$$

has a high coverage (nearly 3,000 known words in the lexicon) and a high precision (99% correct when applied to unknown words). In this case, we added the following rule to the Brill-formatted ruleset:

NN *mente* fhassuf 5 RB x.

In other words, this rule changes the NN tag of nouns that end in *-mente* to RB (adverb).

Table 2. Sample suffixes, with frequency and tag distribution

Suffix	Frequency	Parts of speech
-mente	2,969	99% RB, 0.5% JJ, 0.5% NN
-zione	2,501	99% NN, 0.5% JJ, 0.5% NNP
-abile	1,400	97% JJ, 2% NN, 0.5% RB, 0.5% NNP
-mento	1,375	99% NN, 0.5% VB, 0.5% JJ
-atore	1,28	84% NN, 16% JJ

2.6   Contextual Rules

Ambiguity occurs in most languages. For example in English, in *I can, you can* or *we can, can* is a verb. In *a can* and *the can* it is a noun. We could generalize this in two contextual rules:

- PRP + *can* → VB, and
- DT + *can* → NN.

We automatically constructed a set of 20 contextual rules for Italian derived from the WaCky corpus, using Brill's algorithm. Brill's algorithm takes a tagged corpus, extracts chunks of successive words and tags, and iteratively selects those chunks that increase tagging accuracy. We then proceeded to update and expand this set by hand to 45 rules.

This is a weak step in our approach, since it relies on a previously tagged corpus, which may not exist for other languages. A fully automatic approach would be to look for chunks of words that we know are unambiguous (i.e., one possible tag) and then bootstrap iteratively.

3 Evaluation

We evaluated the tagger (=Wiktionary lexicon + Wikipedia suffix rules + assorted contextual rules) on tagged sentences from the WaCky corpus (1M words). WaCky uses the Tanl tagset, which we mapped to Penn Treebank tags for comparison. Some fine-grained information is lost in the conversion. For example, the Tanl tagset differentiates between demonstrative determiners (DD) and indefinite determiners (DI), which we both mapped to Penn Treebank's DT (determiner). We ignored non-sentence-break punctuation marks; their tags differ between Tanl and Penn Treebank but they are unambiguous. We achieve an overall accuracy of 92.9% (=lexicon 85.8%, morphological rules +6.6%, contextual rules +0.5%).

Accuracy is defined as the percentage of tagged words in WaCky that is tagged identically by our tagger (i.e., for n words, if for word i WaCky says NN and our tag tagger says NN = $+1/n$ accuracy). Table 3 shows an overview of overlap between the tagger and WaCky for the most frequent tags.

Table 3. Breakdown of accuracy for frequent tags

NN	IN	VB	DT	JJ	RB	PRP	CC
(270,000)	(170,000)	(100,000)	(90,000)	(80,000)	(35,000)	(30,000)	(30,000)
94.8%	95.9%	88.5%	90.0%	84.6%	88.6%	82.1%	96.8%

For comparison, we also tested the Italian Perceptron model for OpenNLP and the Italian TreeTagger (Achim Stein's parameter files) against the same WaCky test set. With OpenNLP, we obtain 97.1% accuracy. With TreeTagger, we obtain 83.6% accuracy. This is because TreeTagger does not recognize some proper nouns (NNP) that occur in WaCky and tags some capitalized determiners (DT) such as *La* and *L'* as NN. Both issues would not be hard to address.

We note that our evaluation setup has a potential contamination problem: we used WaCky to obtain a base set of contextual rules (Sect. 2.6), and later on we used the same data for testing. We aim to test against other corpora as they become (freely) available.

4 Related Research

Related work on weakly supervised taggers using Wiktionary has been done by Täckström, Das, Petrov, McDonald and Nivre [17] using Conditional Random Fields (CRFs); by Li, Graça and Taskar [18] using Hidden Markov Models

(HMMs); and by Ding [19] for Chinese. CRF and HMM are statistical machine learning methods that have been successfully applied to natural language processing tasks such as POS tagging.

Täckström et al. and Li et al. both discuss how Wiktionary can be used to construct POS taggers for different languages using bilingual word alignment, but their approaches to infer tags from the available resources differ. We differ from those works in having inferred contextual rules both manually and from a tagged Italian corpus (discussed in Sect. 2.6).

Ding used Wiktionary with Chinese-English word alignment to construct a Chinese POS tagger, and improved the performance of its model with a manually annotated corpus. Hidden Markov Models are used to infer tags.

5 Future Work

We have constructed a simple POS tagger for Italian by mining Wiktionary, Wikipedia and WaCky with weak supervision, contributing to the growing body of work that employs Wiktionary as a useful resource of lexical and semantic information. Our method should require limited effort to be adapted to other languages. It would be sufficient to direct our HTML crawler to another language section on Wiktionary, with small adjustments to the source code. Wiktionary supports, up to now, about 30 languages.

As discussed in Sect. 4, other related research uses different techniques (i.e., HMM), often with better results. In future research we want to compare our approach with those taggers and verify what is the gap between our and other methods.

Finally, Pattern has functionality for sentiment analysis for English, French and Dutch, based on polarity scores for adjectives (see [9]). We are now using our Italian tagger to detect frequent adjectives in product reviews, annotating these adjectives, and expanding Pattern with sentiment analysis for the Italian language.

References

1. Schmid, H.: Probabilistic part-of-speech tagging using decision trees. In: Proceedings of International Conference on New Methods in Language Processing, vol. 12, pp. 44–49 (September 1994)
2. Morton, T., Kottmann, J., Baldridge, J., Bierner, G.: Opennlp: A java-based nlp toolkit (2005)
3. Pianta, E., Zanoli, R.: TagPro: A system for Italian PoS tagging based on SVM. Intelligenza Artificiale 4(2), 8–9 (2007)
4. Tamburini, F.: PoS-tagging Italian texts with CORISTagger. In: Proc. of EVALITA 2009. AI*IA Workshop on Evaluation of NLP and Speech Tools for Italian (2009)
5. Baroni, M., Bernardini, S., Ferraresi, A., Zanchetta, E.: The WaCky wide web: a collection of very large linguistically processed web-crawled corpora. Language Resources and Evaluation 43(3), 209–226 (2009)

6. Attardi, G., Fuschetto, A., Tamberi, F., Simi, M., Vecchi, E.M.: Experiments in tagger combination: arbitrating, guessing, correcting, suggesting. In: Proc. of Workshop Evalita, p. 10 (2009)
7. Søgaard, A.: Ensemble-based POS tagging of Italian. In: The 11th Conference of the Italian Association for Artificial Intelligence, EVALITA, Reggio Emilia, Italy (2009)
8. Dell'Orletta, F.: Ensemble system for Part-of-Speech tagging. In: Proceedings of EVALITA, p. 9 (2009)
9. De Smedt, T., Daelemans, W.: Pattern for Python. The Journal of Machine Learning Research 98888, 2063–2067 (2012)
10. Brill, E.: A simple rule-based part of speech tagger. In: Proceedings of the Workshop on Speech and Natural Language, pp. 112–116. Association for Computational Linguistics (February 1992)
11. Reese, S., Boleda, G., Cuadros, M., Padró, L., Rigau, G.: Wikicorpus: A word-sense disambiguated multilingual Wikipedia corpus (2010)
12. Schneider, G., Volk, M.: Adding manual constraints and lexical look-up to a Brill-tagger for German. In: Proceedings of the ESSLLI 1998 Workshop on Recent Advances in Corpus Annotation, Saarbrücken (1998)
13. Sagot, B.: The Lefff, a freely available and large-coverage morphological and syntactic lexicon for French. In: 7th International Conference on Language Resources and Evaluation, LREC 2010 (2010)
14. Daelemans, W., Zavrel, J., Berck, P., Gillis, S.: MBT: A memory-based part of speech tagger generator. In: Proceedings of the Fourth Workshop on Very Large Corpora, pp. 14–27 (August 1996)
15. Collins, M.: Discriminative training methods for hidden markov models: Theory and experiments with perceptron algorithms. In: Proceedings of the ACL 2002 Conference on Empirical Methods in Natural Language Processing, vol. 10, pp. 1–8. Association for Computational Linguistics (July 2002)
16. Toutanova, K., Klein, D., Manning, C.D., Singer, Y.: Feature-rich part-of-speech tagging with a cyclic dependency network. In: Proceedings of the 2003 Conference of the North American Chapter of the Association for Computational Linguistics on Human Language Technology, vol. 1, pp. 173–180. Association for Computational Linguistics (May 2003)
17. Täckström, O., Das, D., Petrov, S., McDonald, R., Nivre, J.: Token and type constraints for cross-lingual part-of-speech tagging. Transactions of the Association for Computational Linguistics 1, 1–12 (2013)
18. Li, S., Graça, J.V., Taskar, B.: Wiki-ly supervised part-of-speech tagging. In: Proceedings of the 2012 Joint Conference on Empirical Methods in Natural Language Processing and Computational Natural Language Learning, pp. 1389–1398. Association for Computational Linguistics (July 2012)
19. Ding, W.: Weakly supervised part-of-speech tagging for chinese using label propagation (2012)
20. Marcus, M.P., Marcinkiewicz, M.A., Santorini, B.: Building a large annotated corpus of English: The Penn Treebank. Computational Linguistics 19(2), 313–330 (1993)
21. Petrov, S., Das, D., McDonald, R.: A universal part-of-speech tagset.arXiv preprint arXiv:1104 (2011)
22. Collins, A.M., Loftus, E.F.: A spreading-activation theory of semantic processing. Psychological Review 82(6), 407 (1975)

Enhancing Multilingual Biomedical Terminologies via Machine Translation from Parallel Corpora

Johannes Hellrich and Udo Hahn

Jena University Language & Information Engineering (JULIE) Lab,
Friedrich-Schiller-Universität Jena,
Jena, Germany
http://www.julielab.de

Abstract. Creating and maintaining terminologies by human experts is known to be a resource-expensive task. We here report on efforts to computationally support this process by treating term acquisition as a machine translation-guided classification problem capitalizing on parallel multilingual corpora. Experiments are described for French, German, Spanish and Dutch parts of a multilingual biomedical terminology, for which we generated 18k, 23k, 19k and 12k new terms and synonyms, respectively; about one half relate to concepts that have not been lexically labeled before. Based on expert assessment of a sample of the novel German segment about 80% of these newly acquired terms were judged as linguistically correct and bio-medically reasonable additions to the terminology.

1 Introduction

Creating and maintaining terminologies by human experts is known to be a resource-expensive task. The life sciences, medicine and biology, in particular, are one of the most active areas for the development of terminological resources covering a large variety of thematic fields such as species' anatomies, diseases, chemical substances, genes and gene products. These activities have already been bundled in several terminology portals which combine up to some hundreds of these specialized terminologies and ontologies, such as the Unified Medical Language System[1] (UMLS) [1], the Open Biological and Biomedical Ontologies (OBO)[2] [24] and the NCBO BIOPORTAL[3] [30].

For the English language, a stunning variety of broad-coverage terminological resources have been developed over time. Also, many initiatives have been started by non-English language communities to provide human translations of the authoritative English sources, each in the home countries' native languages. Yet despite long-lasting efforts, the term translation gaps encountered between English and the non-English languages remain large—even for prominent and usually well-covered European languages, such as French, Spanish or German. For instance, the non-English counterparts of the UMLS for the languages just mentioned lack between 65 to 94% of the coverage of the English UMLS (see Table 1). Hence, the need for automatic means of terminological acquisition seems obvious.

---

[1] http://www.nlm.nih.gov/research/umls/
[2] http://www.obofoundry.org/
[3] http://bioportal.bioontology.org/

E. Métais, M. Roche, and M. Teisseire (Eds.): NLDB 2014, LNCS 8455, pp. 9–20, 2014.

Table 1. Number of synonyms per language in a slightly pruned UMLS version (for details of the experimental set-up, cf. Section 3) and coverage with respect to the English language

Language	Synonyms	Coverage (w.r.t. English)
English	1,822k	100.0%
Spanish	644k	35.3%
French	137k	7.5%
German	120k	6.6%
Dutch	116k	6.4%

In this paper, we report on efforts to partially close this gap between the English and selected non-English languages (French, German, Spanish and Dutch) through natural language processing technologies, basically by applying statistical machine translation (SMT) [14] and named entity recognition (NER) techniques [19] to selected parallel corpora [26]. Parallel corpora are a particularly rich source for SMT because they contain documents which are direct translations of each other. Since these translations are supplied by human experts, they are generally considered to be of a high degree of translation quality. SMT systems build on these raw data in order to generate statistical translation models from the distribution and co-occurrence statistics of single words in the monolingual document sets together with word and phrase alignments between paired bilingual document sets.

For the experiments discussed below, we employed two different types of parallel corpora, namely MEDLINE titles and EMEA drug leaflets. MEDLINE[4] supplies English translations of titles from national non-English scientific journals (especially numerous for Spanish, French and German). Similarly, the EMEA corpus[5] [25] provides drug information in 22 languages, e.g. about one million sentences each for Spanish, French, German and Portuguese. All these documents have been carefully crafted by the authors of journal articles or educated support staff such as editors or translators. From these resources, we automatically acquired for the French, German, Spanish and Dutch parts of the UMLS 18k, 23k, 19k and 12k new entries, respectively. Based on expert assessment of a sample of the new German segment about 80% of these new terms were judged as linguistically correct and bio-medically plausible terminology enhancements.

2 Related Work

SMT uses a statistical model to map expressions from a source language into expressions of a target language. Nowadays most SMT models are phrase-based, i.e. they use the probability of two text segments being translational equivalents and the likelihood of a text segment existing in the target language to model translations [16]. SMT model building requires an ample amount of training material, usually parallel corpora. In the meantime, open source systems such as MOSES [15] have been shown to generate high-quality translations for corpus-wise well-resourced languages such as German, Spanish

---

[4] http://mbr.nlm.nih.gov/Download/
[5] http://opus.lingfil.uu.se/EMEA.php

or French [31]. Notable efforts to use parallel corpora for terminology translation are due to Déjean et al. [5] and Deléger et al. [6,7], with German and French as the respective target languages.

Since parallel corpora are often rare (for some domains even non-existing) and limited in scope, researchers have been looking for alternatives. One way to go are comparable corpora. Just as parallel corpora they consist of texts from the same domain, but unlike parallel corpora there exists no 1:1 mapping between content-related sentences in the different languages. Comparable corpora are larger in size and easier to collect [22], yet they provide less precise results. Thus they are sometimes used in combination with parallel corpora, leading to better results than each on its own [5].

Comparable corpora are typically exploited with context-based projection methods, i.e. context vectors are collected for words in both languages which get partially translated with a seed terminology—words with similar vectors are thus likely to be translations of each other. A short overview of the use of comparable corpora in SMT and the influence of different parameters is provided by Laroche & Langlais [17]. Skadiņa et al. [23] show how a subset can be extracted from a comparable corpus which can be treated as a parallel corpus. Alternatively, subwords (i.e. morpheme-like segments [11]) with known translations (e.g. 'cyto' :: 'cell') can be used to identify new terms composed from these elements [8].

As another alternative, non-English terminology can also be harvested from documents in a monolingual fashion, without an a priori reference anchor in English (hence, no parallel corpora are needed). These approaches are primarily useful if no initial multilingual terminology exists. After extraction, these terminologies need to be aligned either employing purely statistical or linguistically motivated similarity criteria [4,28].

In general, terminology extraction in the bio-medical field is a tricky task because the new terms to extract do not only appear as single words (e.g. 'appendicitis') but much more so as multiword expressions (e.g. 'Alzheimer's disease' or 'acquired immunodeficiency syndrome'). Approaches towards finding these expressions can be classified as either pattern-based, using e.g. manually created POS patterns that signal termhood, or statistically motivated, utilizing e.g. phrase alignment techniques. Pattern-based approaches such as the ones described by Déjean et al. [5] or Bouamor et al. [3] are fairly common. The downside of these approaches is their need for POS patterns, which are often hand-crafted and may become cumbersome to read and write. Alternative approaches use statistical data, like the translation probabilities of the single words of a term (treated as a bag of words) [27] or some kind of phrases. These can either be linguistically motivated, i.e. use POS information [18] or be purely statistical and derived from the model produced by a phrase-based SMT system.

Also in terms of evaluation, a large diversity of approaches can be observed. Some groups report only precision data based on the number of correct translations produced by their system [6]. Others issue F-scores based on the system's ability to reproduce a (sample) terminology [5]. In these studies, numbers for new and correct translations range between 62% and 81% [6,7]. Unfortunately, these results are not only grounded in the systems' performance, but are rather strongly influenced by both the chosen terminology and the parallel corpora being used—thus comparisons between systems are hampered by the lack of a common methodology and standard for evaluation.

3 Experimental Set-Up

Methodologically speaking, our approach primarily combines off-the-shelf components, namely the LINGPIPE[6] gazetteer for lexical processing, GIZA++ and MOSES[7] [20,15] for generating a phrase-based statistical machine translation model, JCORE, a Conditional Random Field-based system for biomedical named entity recognition (used to indicate termhood of phrases) [10], and WEKA[8] [12] as the machine learning repository from which we took a Maximum Entropy classifier to combine NER and SMT information.

We distinguish three steps in our experimental set-up: corpus set-up, candidate generation and candidate filtering. Both the corpora and the UMLS version we used were taken from the material provided for the CLEF-ER 2013 challenge[9] [21], where a variant of the pipeline described below was used to extract new terms for annotating JULIE Lab's challenge contribution [13].

3.1 Corpus Set-Up

We started by merging the MEDLINE and EMEA parallel corpora, resulting in one corpus per language pair. These corpora contained 860k, 713k, 389k and 195k parallel sentences for the German, French, Spanish and Dutch language, respectively (see Table 2 for details).

Table 2. Number of MEDLINE titles and EMEA sentences in the parallel corpus for each language. These numbers reflect the final version of the CLEF-ER material and are more recent than the overview on the challenge website.

Language	MEDLINE titles	EMEA sentences
German	719k	141k
French	572k	141k
Spanish	248k	141k
Dutch	54k	141k

Next, a flat annotation for already known 'biomedical entities' was performed, i.e. without any further distinction of more fine-grained semantic groups, such as *Disease* or *Anatomy* (which is common in the UMLS framework). This step was based on the UMLS-derived terminology using a LINGPIPE-based gazetteer, i.e. each textual item was checked whether it could be matched with an entry in that terminology. 10% of each of the annotated corpora were set aside and subsequently used to train language-specific JCORE NER systems in order to find biomedical entities using ML models. Thus the probabilities provided by the systems characterize biomedical termhood of a

---

[6] http://alias-i.com/lingpipe/
[7] http://www.statmt.org/moses/
[8] http://www.cs.waikato.ac.nz/~ml/weka/
[9] https://sites.google.com/site/mantraeu/clef-er-challenge

Table 3. Performance of the NER system using 10-fold cross-validation for each language

Language	F_1-score	Precision	Recall
Spanish	0.82	0.87	0.79
German	0.78	0.88	0.70
French	0.77	0.83	0.72
Dutch	0.58	0.92	0.43

word (sequence) under scrutiny. Performance data for these systems can be found in Table 3. The remaining 90% of the corpora were used to train an SMT model with GIZA++ and MOSES.

3.2   Candidate Generation

The so-called phrase table of the SMT model trained this way contains phrase pairs, such as *'an indication of tubal cancer'* :: *'als Hinweis auf ein Tubenkarzinom'* and translation probabilities for each pair—conditional probabilities for a phrase in the target language (either German, French, Spanish or Dutch) and a phrase in the source language (English) being translations of each other. Term candidates for enriching the UMLS were produced by selecting those phrase pairs translating a known English biomedical term into one of the target languages.

3.3   Candidate Filtering

A naive *baseline* system could accept all those candidates from the phrase table, yet most of them would presumably be wrong. Since the phrases in our system are statistically rather than linguistically motivated, it simply maps text segments onto each other, producing scruffy translations like *'cancer'* :: *'Krebs zu' ['cancer to']*, which contain not only the relevant biomedical term (*'Krebs' ['cancer']*) but also (sometimes multiple) non-relevant text tokens (here, *'zu' ['to']*).

A first refinement keeps only those candidates which are most probable according to the SMT model, as those will be both relatively frequent and specific. This can be done by using a *cut-off* which selects the n most likely term candidates based on one of the aforementioned translational probabilities, the direct phrase translation probability (selected through a pre-test). Thus for each English biomedical term in the corpus the n most probable translations are produced as new biomedical terms for the target language.

Further filtering was carried out by selecting candidates with a *classifier*. We trained a WEKA-based Maximum Entropy classifier by using the gazetteer-based annotations generated in Step 3.1 as training material and tested the following features and their combination:

– **Phrase translation probabilities.** Translation probabilities from the SMT model, as a more refined alternative to the aforementioned cut-off.

- **Named entity recognition (NER).** JCORE's judgments on each candidate's term-hood. The NER system was run on all sentences containing the candidate, and averaging the probabilities (taking '0', if no match was found).
- **Length.** The smoothed ratio between the respective lengths of the term candidate and its translation equivalent (to exclude overly length-biased and thus unlikely translation pairs).

4 Results and Evaluation

In general, three types of translation equivalents can be found for the non-English terminologies using our methodological approach: *known terms* and *additional synonyms* for concepts already labeled in a terminology, as well as *entirely new terms* for concepts with no prior label in a terminology. To evaluate the extracted terms we used both an automatic evaluation procedure and human expert assessments, dealing with already known terms plus additional synonyms, and entirely new terms, respectively.

4.1 Automatic Evaluation

In the first setting, we measured the system's ability to re-invent UMLS fragments, using 10-fold cross-validation for the classifier-based systems. This evaluation was carried out concept-wise, i.e. a term being a synonym for multiple concepts, or a translation thereof, was examined multiple times depending on the number of synonyms we encountered. A term was counted as being *correct*, if it was contained in the set of synonyms in the respective language for the UMLS concept, e.g. *'Krebs'* would be a correct German synonym for the concept *Neoplasm malignant*, i.e. *Cancer*. False negatives (for recall calculation) were counted based on an upper bound estimate, namely by taking those concepts into account for which the gazetteer system had annotated terms in two aligned sentences; we call those items *traceable* (see Table 4 for a overview). Thus, the concept *Neoplasm malignant* is only traceable, if the corpus contains a sentence pair like:

'He suffers from <u>cancer</u>.' — *'Er leidet an <u>Krebs</u>.'*

Table 4. Number of traceable terms and synonyms, i.e. known terms and synonyms in parallel sentences, per language, used as an upper bound to calculate recall

Language	Traceables
Spanish	19,797
French	17,492
German	15,428
Dutch	2,017

Unfortunately, this approach erroneously counts additional synonyms and entirely new terms as incorrect, since they are not yet contained in the terminology. This drawback may also hamper training, as valid additions are treated as negative examples during training.

Still, the automatic analysis allows to assess the influence of different approaches and feature combinations on term extraction for all four languages (results are listed in Table 5). A system without any candidate classification, i.e. merely enriching the terminology with all translations contained in the SMT model, is listed as a naive *baseline* and trivially pushes recall. We also list three *cut-off* based system discarding all but the most probable n translations of an English term; they achieve intermediate F_1-scores, with the topmost choice always leading to the best F-score, and have recall values comparable to the baseline, indicating a high capability of the SMT system to identify relevant phrases. Finally, we list the classifier-based systems using different feature combinations as noted in Table 5.

Table 5. Evaluation of the re-invention performance by considering the UMLS as a gold standard, with F1-score (F), Precision (P) and Recall (R) for all four languages. Listing a naive baseline (without candidate filtering), cut-off selection systems picking up the top n candidates (by direct phrase translation probability) for each term and classifier-based systems with different feature combinations referring to the conditions described in Section 3.3.

Method	German			French			Spanish			Dutch		
	F	P	R	F	P	R	F	P	R	F	P	R
Baseline	0.14	0.07	**0.98**	0.13	0.07	**0.98**	0.19	0.11	**0.97**	0.11	0.06	**0.95**
cut-off, top 1	0.40	0.25	0.96	0.37	0.23	0.97	0.48	0.32	0.95	0.39	0.25	0.91
cut-off, top 3	0.22	0.12	**0.98**	0.22	0.13	**0.98**	0.28	0.16	**0.97**	0.19	0.10	0.94
cut-off, top 5	0.19	0.18	**0.98**	0.18	0.10	**0.98**	0.25	0.14	**0.97**	0.15	0.08	**0.95**
Classifier 1 (only NER)	0.18	0.36	0.13	0.31	0.50	0.22	0.28	0.48	0.21	0.14	0.29	0.10
Classifier 2 (only phrase translation probability)	0.62	0.55	0.72	0.54	0.55	0.54	0.72	0.63	0.84	0.55	0.40	0.84
Classifier 3 (Classifier 2 + NER)	0.65	**0.61**	0.70	0.59	0.57	0.61	0.72	0.62	0.86	0.54	0.40	0.84
Classifier 4 (Classifier 3 + length)	**0.67**	0.60	0.75	**0.62**	**0.60**	0.63	**0.74**	**0.65**	0.87	**0.56**	**0.41**	0.88

The classifier version with all tested features was the best-performing system by F_1-scores over all languages. The classifier using only NER information and no translation probabilities (i.e. accepting any translation proposed by the SMT system, if it was deemed to be some kind of biomedical entity) performed under par, whereas a system using only translation probabilities performs close to the overall best system, indicating again the high quality of the phrases generated by the SMT system. The effect of the NER feature on the performance of classifiers 3 and 4 seems to be strongly language-dependent, yet there is no clear connection to the F_1-scores reported in Table 3, with Spanish and Dutch profiting equally little despite being the best and worst, respectively, in terms of NER performance. Recall for all classifier systems, except the phrase-translation-probability-only classifier 2, is rather high. This can be explained by the fact that all systems filter on the baseline, which provides recall values in the high nineties.

While performance is comparable over all languages, Dutch suffers from precision problems while achieving above-average recall—this anomaly could be explained both by the low number of traceable Dutch terms (i.e. the ceiling for recall calculation and positive examples for classifier training; see Table 4) and the smaller corpus size (see Table 2). The former seems more likely, as Spanish achieves the overall best F_1-score and precision values despite a medium corpus size, yet has the highest number of traceable terms.

4.2 Expert Judgment

The classifier-based terminology acquisition system with all tested features found overall several thousands of entirely new terms and additional synonyms for each of the four languages (see Table 6). Numbers are quite similar for German and French, whereas the much smaller size of the Dutch corpus leads to a higher amount of entirely new terms and the inverse seems to be true for Spanish due to the comparably larger coverage of the Spanish UMLS.

Table 6. Number of terms extracted by the classifier system with all features for all four languages. We distinguish three types of extracted items: already known terms and synonyms are already contained as a label for this concept and language (term re-invention), additional synonyms supplement terms/synonyms known for this concept and language by new ones (term enrichment), entirely new terms comprise really novel terms for a concept in that language previously unlabeled (term learning).

Language	Already known terms & synonyms	Additional synonyms	Entirely new terms
German	11k	13k	10k
French	10k	11k	7k
Spanish	19k	17k	2k
Dutch	4k	6k	6k

The terms and synonyms not yet contained in the UMLS were assessed by two experts, both bioinformatics graduate students, utilizing online medical dictionaries. Their judgments were collected based on a sample of 100 extracted German terms. 80% of these were unanimously considered as being correct and reasonable from a domain knowledge perspective. A surprisingly big chunk of these (49%) is spelled identically as a known English term for the same concept, e.g. the name of the tuna species *'Thunnus thynnus'*. This is caused by a large number of entries listed as English in the UMLS, despite the fact that they are used worldwide (often due to their Latin origin) to denote the same concept. Furthermore, 8% are linguistically valid translations of an English term, although they are not yet suited as terminology entries. This is most often the case for inflected forms, e.g. *'linken Nebenniere'* instead of *'linke Nebenniere'* as a synonym of *'entire left adrenal gland'*. Only a small amount (6%) of the errors of our system raise serious concerns as mistranslations, probably caused by partial matches during phrasal alignment, e.g. *'Flussmessung'* [*'flow measurement'*] as a synonym for *'cardiac flow'*. Whereas the latter are probably hard to avoid, the former should be eliminated by incorporating a morphological stemmer during candidate generation. Finally another 6% of the terms were accepted by only one expert, e.g. *'Phagedänismus''* as a synonym for *'Phagedenic ulcer'*.

5 Conclusions

The terminology acquisition system for French, German, Spanish and Dutch parts of a biomedical UMLS-derived terminology yielded 18k, 23k, 19k and 12k additional synonyms (for terms already listed in the terminology) and entirely new terms (for concepts lacking any official label in this language), respectively. About half of these are really new (e.g. not just Latin terms occurring in many languages) and thus of particularly high value for terminological applications. The system is also good at discriminating domain-specific terms given evidence from a German sample which revealed that 80% of the extracted terms are correct and reasonable for this domain. Upon completion of our work, the extracted terms will be made publicly available to the relevant national and terminological authorities.

Using texts from other sources, like web pages [22] or WIKIPEDIA [2], including the use of comparable corpora [4,23], should be straightforward alternatives to further develop our system and decouple it from its reliance on strictly parallel corpora. We also plan to incorporate alternative classifiers (such as SVMs) with an enriched feature set, in particular, including 'phrasal' types of features such as term collocations [29] and additional indicators of termhood such as term extraction metrics [9]. Adding more features, e.g. morphological clues or information about collocations, could also lead to improved performance. To prevent the extraction of unwarranted terms we plan to incorporate more sophisticated stemming functionality into the system so that inflected forms can easily be removed. While the system is basically language and domain-independent, both candidate generation and candidate classification depend on resources, i.e. parallel corpora and existing multilingual terminologies, which may not be available for all use cases.

Acknowledgments. This work is funded by the European Commission's 7th Framework Programme for small or medium-scale focused research actions (STREP) from the Information Content Technologies Call FP7-ICT-2011-4.1, Challenge 4: Technologies for Digital Content and Languages, Grant No. 296410.

References

1. Bodenreider, O.: The Unified Medical Language System (Umls): Integrating biomedical terminology. Nucleic Acids Research 32(Database issue), D267–D270 (2004)

2. Bouamor, D., Popescu, A., Semmar, N., Zweigenbaum, P.: Building specialized bilingual lexicons using large-scale background knowledge. In: EMNLP 2013 – Proceedings of the 2013 Conference on Empirical Methods in Natural Language Processing. A meeting of SIG-DAT, a Special Interest Group of the ACL, Seattle, WA, USA, October 18-21, pp. 479–489. Association for Computational Linguistics, ACL (2013)

3. Bouamor, D., Semmar, N., Zweigenbaum, P.: Identifying bilingual multi-word expressions for statistical machine translation. In: LREC 2012 – Proceedings of the 8th International Conference on Language Resources and Evaluation, Istanbul, Turkey, May 23-25, pp. 674–679. European Language Resources Association (ELRA, Paris (2012)

4. Ştefănescu, D.: Mining for term translations in comparable corpora. In: BUCC 5 – Proceedings of the 5th Workshop on Building and Using Comparable Corpora: Language Resources for Machine Translation in Less-Resourced Languages and Domains @ LREC 2012: 8th International Conference on Language Resources and Evaluation, Istanbul, Turkey, pp. 98–103. European Language Resources Association (ELRA, Paris (2012)

5. Déjean, H., Gaussier, E., Renders, J.M., Sadat, F.: Automatic processing of multilingual medical terminology: Applications to thesaurus enrichment and cross-language information retrieval. Artificial Intelligence in Medicine 33(2), 111–124 (2005)

6. Deléger, L., Merkel, M., Zweigenbaum, P.: Enriching medical terminologies: An approach based on aligned corpora. In: Hasman, A., Haux, R., van der Lei, J., De Clercq, E., Roger France, F.H. (eds.) MIE 2006 – Proceedings of the 20th International Congress of the European Federation for Medical Informatics, Maastricht, The Netherlands, August 27-30. Studies in Health Technology and Informatics, vol. 124, pp. 747–752. IOS Press, Amsterdam (2006)

7. Deléger, L., Merkel, M., Zweigenbaum, P.: Translating medical terminologies through word alignment in parallel text corpora. Journal of Biomedical Informatics 42(4), 692–701 (2009)

8. Delpech, E., Daille, B., Morin, E., Lemaire, C.: Extraction of domain-specific bilingual lexicon from comparable corpora: Compositional translation and ranking. In: COLING 2012 – Proceedings of the 24th International Conference on Computational Linguistics: Technical Papers, Mumbai, India, December 8-15, pp. 745–762. Indian Institute of Technology (2012)

9. Frantzi, K.T., Ananiadou, S., Mima, H.: Automatic recognition of multi-word terms: The C-value/NC-value method. International Journal on Digital Libraries 3(2), 115–130 (2000)

10. Hahn, U., Buyko, E., Landefeld, R., Mühlhausen, M., Poprat, M., Tomanek, K., Wermter, J.: An overview of JCoRe, the Julie Lab Uima component repository. In: Proceedings of the LREC 2008 Workshop 'Towards Enhanced Interoperability for Large HLT Systems: UIMA for NLP', Marrakech, Morocco, pp. 1–7. European Language Resources Association (ELRA, Paris (2008)

11. Hahn, U., Markó, K.G., Schulz, S.: Subword clusters as light-weight interlingua for multilingual document retrieval. In: MT Summit X – Proceedings of the 10th Machine Translation Summit of the International Association for Machine Translation, Phuket, Thailand, September 12-16, pp. 17–24. Asia-Pacific Association for Machine Translation, AAMT (2005)

12. Hall, M., Frank, E., Holmes, G., Pfahringer, B., Reutemann, P., Witten, I.H.: The Weka data mining software: An update. ACM SIGKDD Explorations 11(1), 10–18 (2009)

13. Hellrich, J., Hahn, U.: The julie Lab mantra system for the clef-er 2013 challenge. In: CLEF 2012, CLEF 2013 Evaluation Labs and Workshop Online Working Notes, Valencia, Spain (September 25, 2013),
http://www.clef-initiative.eu/documents/71612/a132d6c9-b0f1-48a4-a0c5-648e5127e229

14. Koehn, P.: Statistical Machine Translation. Cambridge University Press, Cambridge (2010)

15. Koehn, P., Hoang, H., Birch, A., Callison-Burch, C., Federico, M., Bertoldi, N., Cowan, B., Shen, W., Moran, C., Zens, R., Dyer, C., Bojar, O., Constantin, A., Herbst, E.: Moses: Open source toolkit for statistical machine translation. In: ACL 2007 – Proceedings of the 45th Annual Meeting of the Association for Computational Linguistics, Prague, Czech Republic, June 25-27. Proceedings of the Interactive Poster and Demonstration Sessions, vol. Companion, pp. 177–180. Association for Computational Linguistics, ACL (2007)
16. Koehn, P., Och, F.J., Marcu, D.: Statistical phrase-based translation. In: HLT-NAACL 2003 – Human Language Technology Conference of the North American Chapter of the Association for Computational Linguistics, Edmonton, Canada, May 27-June 1, vol. 1, pp. 48–54. Association for Computational Linguistics (ACL), Stroudsburg (2003)
17. Laroche, A., Langlais, P.: Revisiting context-based projection methods for term-translation spotting in comparable corpora. In: COLING 2010 – Proceedings of the 23rd International Conference on Computational Linguistics, Beijing, China, August 23-27, pp. 617–625. Tsinghua University Press, Beijing (2010)
18. Lefever, E., Macken, L., Hoste, V.: Language-independent bilingual terminology extraction from a multilingual parallel corpus. In: EACL 2009 – Proceedings of the 12th Conference of the European Chapter of the Association for Computational Linguistics, Athens, Greece, March 30-April 3, pp. 496–504. Association for Computational Linguistics (2009)
19. Nadeau, D., Sekine, S.: A survey of named entity recognition and classification. Linguisticae Investigationes 30(1), 3–26 (2007)
20. Och, F.J., Ney, H.: A systematic comparison of various statistical alignment models. Computational Linguistics 29(1), 19–51 (2003)
21. Rebholz-Schuhmann, D., et al.: Entity recognition in parallel multi-lingual biomedical corpora: The Clef-ER Laboratory overview. In: Forner, P., Müller, H., Paredes, R., Rosso, P., Stein, B. (eds.) CLEF 2013. LNCS, vol. 8138, pp. 353–367. Springer, Heidelberg (2013)
22. Resnik, P., Smith, N.A.: The Web as a parallel corpus. Computational Linguistics 29(3), 349–380 (2003)
23. Skadiņa, I., Aker, A., Mastropavlos, N., Su, F., Tufiş, D., Verlic, M., Vasiļjevs, A., Babych, B., Clough, P., Gaizauskas, R., Glaros, N., Lestari Paramita, M., Pinnis, M.: Collecting and using comparable corpora for statistical machine translation. In: LREC 2012 – Proceedings of the 8th International Conference on Language Resources and Evaluation, Istanbul, Turkey, May 23-25, pp. 438–445. European Language Resources Association (ELRA, Paris (2012)
24. Smith, B., Ashburner, M., Rosse, C., Bard, J., Bug, W., Ceusters, W., Goldberg, L., Eilbeck, K., Ireland, A., Mungall, C.J., Leontis, N., Rocca-Serra, P., Ruttenberg, A., Sansone, S.A., Scheuermann, R.H., Shah, N.H., Whetzel, P.L., Lewis, S.E.: The Obo Foundry: Coordinated evolution of ontologies to support biomedical data integration. Nature Biotechnology 25(11), 1251–1255 (2007)
25. Tiedemann, J.: News from Opus: A collection of multilingual parallel corpora with tools and interfaces. In: Nicolov, N., Angelova, G., Mitkov, R. (eds.) RANLP 2009 – Recent Advances in Natural Language Processing. No. 309 in Current Issues in Linguistic Theory, vol. V, pp. 237–248. John Benjamins, Amsterdam (2009)
26. Véronis, J.: From the Rosetta stone to the information society. A survey of parallel text processing. In: Véronis, J. (ed.) Parallel Text Processing. Alignment and Use of Translation Corpora. No. 13 in Text, Speech and Language Technology, pp. 1–24. Kluwer Academic Publ., Dordrecht (2000)
27. Vintar, Š.: Bilingual term recognition revisited: The bag-of-equivalents term alignment approach and its evaluation. Terminology 16(2), 141–158 (2010)
28. Weller, M., Gojun, A., Heid, U., Daille, B., Harastani, R.: Simple methods for dealing with term variation and term alignment. In: TIA 2011 – Proceedings of the 9th International Conference on Terminology and Artificial Intelligence, Paris, France, November 8-10, pp. 87–93 (2011)

29. Wermter, J., Hahn, U.: Paradigmatic modifiability statistics for the extraction of of complex multi-word terms. In: HLT/EMNLP 2005 – Proceedings of the Human Language Technology Conference and the Conference on Empirical Methods in Natural Language Processing, Vancouver, BC, Canada, October 6-8, pp. 843–850. Association for Computational Linguistics (ACL), East Stroudsburg (2005)

30. Whetzel, P.L., Noy, N.F., Shah, N.H., Alexander, P.R., Nyulas, C., Tudorache, T., Musen, M.: BioPortal: Enhanced functionality via new Web services from the National Center for Biomedical Ontology to access and use ontologies in software applications. Nucleic Acids Research 39(Web Server issue), W541–W545 (2011)

31. Wu, C., Xia, F., Deléger, L., Solti, I.: Statistical machine translation for biomedical text: Are we there yet? In: AMIA 2011 – Proceedings of the Annual Symposium of the American Medical Informatics Association. Improving Health: Informatics and IT Changing the World, Washington, DC, USA, October 22-26, pp. 1290–1299. American Medical Informatics Association (2011)

A Distributional Semantics Approach for Selective Reasoning on Commonsense Graph Knowledge Bases

André Freitas[1], João Carlos Pereira da Silva[1,2], Edward Curry[1], and Paul Buitelaar[1,3]

[1] Insight Centre for Data Analytics, National University of Ireland, Galway
[2] Computer Science Department, Federal University of Rio de Janeiro
[3] College of Graduate Studies, University of South Africa

Abstract. Tasks such as question answering and semantic search are dependent on the ability of querying & reasoning over large-scale commonsense knowledge bases (KBs). However, dealing with commonsense data demands coping with problems such as the increase in schema complexity, semantic inconsistency, incompleteness and scalability. This paper proposes a selective graph navigation mechanism based on a distributional relational semantic model which can be applied to querying & reasoning over heterogeneous knowledge bases (KBs). The approach can be used for approximative reasoning, querying and associational knowledge discovery. In this paper we focus on commonsense reasoning as the main motivational scenario for the approach. The approach focuses on addressing the following problems: (i) providing a semantic selection mechanism for facts which are relevant and meaningful in a specific reasoning & querying context and (ii) allowing coping with information incompleteness in large KBs. The approach is evaluated using ConceptNet as a commonsense KB, and achieved *high selectivity, high scalability* and *high accuracy in the selection of meaningful navigational paths*. Distributional semantics is also used as a principled mechanism to cope with information incompleteness.

1 Introduction

Building intelligent applications and addressing simple computational semantic tasks demand coping with large-scale commonsense Knowledge Bases (KBs). Querying and reasoning (Q&R) over large commonsense KBs are fundamental operations for tasks such as Question Answering, Semantic Search and Knowledge Discovery. However, in an open domain scenario, the scale of KBs and the number of direct and indirect associations between elements in the KB can make Q&R grow unmanageable. To the complexity of querying and reasoning over such large-scale KBs, it is possible to add the barriers involved in building KBs with the necessary consistency and completeness requirements.

With the evolution of open data, better information extraction frameworks and crowd-sourcing tools, large-scale structured KBs are becoming more available. This data can be used to provide commonsense knowledge for semantic applications. However, querying and reasoning over this data demands approaches which are able to cope with large-scale, semantically heterogeneous and incomplete KBs.

E. Métais, M. Roche, and M. Teisseire (Eds.): NLDB 2014, LNCS 8455, pp. 21–32, 2014.

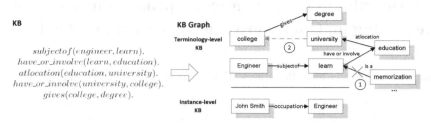

Fig. 1. (1) Selection of meaningful paths, (2) Coping with information incompleteness

As a motivational scenario, suppose we have a KB with the following fact: *'John Smith is an engineer'* and suppose the query *'Does John Smith have a degree?'* is issued over the KB. A complete KB would have the rule *'Every engineer has a degree'*, which would materialize *'John Smith has a degree'*. For large-scale and open domain commonsense reasoning scenarios, model completeness and full materialization cannot be assumed. In this case the information can be embedded in other facts in the KB (Figure 1). The example sequence of relations between *engineer* and *degree* defines a path in a large-scale graph of relations between predicates, which is depicted in Figure 1.

In a large-scale KB, full reasoning can become unfeasible. A commonsense KB would contain vast amounts of facts and a complete inference over the entire KB would not scale to its size. Furthermore, while the example path is a meaningful sequence of associations for answering the example query, there is a large number of paths which are not meaningful under a specific query context. In Figure 1(1), for example, the reasoning path which goes through (1) is not related to the goal of the query (the relation between *engineer* and *degree*) and should be eliminated. Ideally a query and reasoning mechanism should be able to filter out facts and rules which are unrelated to the Q&R context. The ability to select the minimum set of facts which should be applied in order to answer a specific user information need is a fundamental element for enabling reasoning capabilities for large-scale commonsense knowledge bases.

Additionally, since information completeness of the KBs cannot be guaranteed, one missing fact in the KB would be sufficient to block the reasoning process. In Figure 1(2) the lack of a fact connecting university and college eliminates the possibility of answering the query. Ideally Q&R mechanisms should be able to cope with some level of KB incompleteness, approximating and filling the gaps in the KBs.

This work proposes a *selective reasoning approach* which uses a *hybrid distributional-relational semantic model* to address the problems previously described. Distributional semantic models (DSMs) use statistical co-occurrence patterns, automatically extracted from large unstructured text corpora, to support the creation of comprehensive quantitative semantic models. In this work, DSMs are used as complementary semantic layer to the relational model, which supports coping with semantic approximation and incompleteness. The proposed approach focuses on the following contributions:

– provision of a selective Q&R approach using a distributional semantics heuristics, which reduces the search space for large-scale KBs at the same time it maximizes paths which are more meaningful for a given reasoning context;

– definition of a Q&R model which copes with the information incompleteness present at the KB, using the distributional model to support semantic approximations, which can fill the lack of information in the KB during the reasoning process;

This work is organized as follows: section 2 provides an introduction on distributional semantics; section 3 describes the τ-Space distributional-relational semantic model which is used for the selection reasoning mechanism; section 4 describes the selective reasoning mechanism (*distributional navigational algorithm*); section 5 provides an evaluation of the approach using Explicit Semantic Analysis (ESA) as a distributional semantic model and ConceptNet [11] as KB; section 6 describes related work and finally, section 7 provides conclusions and future work.

2   Distributional Semantics

In this work *distributional semantics* supports the definition of an *approximative semantic navigational approach* in a knowledge base, where the graph concepts and relations are mapped to vectors in a *distributional vector space*.

Distributional semantics is defined upon the assumption that the context surrounding a given word in a text provides important information about its meaning [12]. It focuses on the construction of a semantic model for a word based on the statistical distribution of co-located words in texts. These semantic models are naturally represented by Vector Space Models (VSMs), where the meaning of a word can be defined by a weighted vector, which represents the association pattern of co-occurring words in a corpus.

The existence of large amounts of unstructured text on the Web brings the potential to create comprehensive distributional semantic models (DSMs). DSMs can be automatically built from large corpora, not requiring manual intervention on the creation of the semantic model. Additionally, its natural association with VSMs, which are supported by dimensional reduction approaches or data structures such as inverted list indexes can provide a scalability benefit for the instantiation of these models.

The computation of *semantic relatedness measure* between words is one instance in which the strength of distributional models and methods is empirically supported ([3];[2]). The computation of the *semantic relatedness measure* is at the center of this work and it is used as a *semantic heuristics* to navigate in the KB graph, *where the distributional knowledge extracted from unstructured text is used as a general-purpose large-scale commonsense KB, which complements the knowledge present at the relational KB.*

3 τ-Space

The τ-*Space* [1] is a *distributional structured vector space model* which allows the representation of the elements of a graph KB under the grounding of a distributional semantic model. This work improves the formalisation on the definition of the τ-*Space*.

τ-*Space* is built from a *reference corpus* $RC = (Term, Context)$ formed by a set of terms $Term = \{k_1, \cdots, k_t\}$ and a set of context windows $Context = \{c_1, \cdots, c_t\}$.

The set $Term$ is used to define the basis $Term_{basis} = \{\vec{k}_1, \cdots, \vec{k}_t\}$ of unit vectors that spans the *term vector space* VS^{term}.

A context window c_j is represented in VS^{term} as:

$$\vec{c}_j = \sum_{i=1}^{t} v_{i,j} \vec{k}_i \tag{1}$$

where $v_{i,j}$ is 1 if term k_i appears in context window c_j and 0 otherwise.

Analogously, the set of context windows $Context$ is used to define the basis $Context_{basis} = \{\vec{c}_1, \cdots, \vec{c}_t\}$ of vectors that spans the *distributional vector space* VS^{dist}. A given term x is represented in VS^{dist} as:

$$\vec{x} = \sum_{j=1}^{t} w_j \vec{c}_j \tag{2}$$

such that

$$w_j = tf_j \times idf = \frac{freq_j}{count(c_j)} \times \log \frac{N}{n_{c_j}} \tag{3}$$

where w_j is the product of the normalized term frequency tf_j (the ratio between the frequency of term x in the context window c_j and the number of terms inside c_j) and the inverse document frequency idf for the term x (the logarithm of the ratio of the total number of N context windows in the reference corpus RC and the number n_{c_j} of context containing the term x).

Thus, the set of context windows where a term occurs define the concept vectors associated with the term, which is a representation of its meaning on the reference corpus.

4   Embedding the Commonsense KB into the τ-Space

We consider that a commonsense knowledge base KB is formed by a set of *concepts* $\{v_1, \cdots, v_n\}$ and a set of *relations* $\{r_1, \cdots, r_m\}$ between these concepts, both represented as words or short phrases in natural language. Formally, a commonsense knowledge base KB is defined by a *labeled digraph* $G_{KB}^{label} = (V, R, E)$, where $V = \{v_1, \cdots, v_n\}$ is a set of nodes, $R = \{r_1, \cdots, r_m\}$ is a set of relations and E is a set of directed edges (v_i, v_j) labeled with relation $r \in R$ and denoted by (v_i, r, v_j). Alternatively, we can simplify the representation of the KB ignoring their relation labels: Let KB be commonsense knowledge base and $G_{KB}^{label} = (V, R, E)$ be its labeled digraph representation. A simplified representation of KB is defined by a *digraph* $G_{KB} = (V', E')$, where $V' = V$ and $E' = \{(v_i, v_j) : (v_i, r, v_j) \in E\}$. Given the (labeled) graph representation of KB, we have to embed it into the τ-Space. To do that we have to translate the nodes and edges of the graph representation of KB into a vector representation in VS^{dist}. The vector representation of $G_{KB}^{label} = (V, R, E)$ in VS^{dist} is $\vec{G}_{KB_{dist}}^{label} = (\vec{V}_{dist}, \vec{R}_{dist}, \vec{E}_{dist})$ such that:

$$\vec{V}_{dist} = \{\vec{v} : \vec{v} = \sum_{i=1}^{t} u_i^v \vec{c}_i, \text{ for each } v \in V\} \tag{4}$$

$$\overrightarrow{\mathbf{R}}_{dist} = \{\overrightarrow{\mathbf{r}} : \overrightarrow{\mathbf{r}} = \sum_{i=1}^{t} u_i^r \overrightarrow{\mathbf{c}}_i, \text{ for each } r \in R\} \tag{5}$$

$$\overrightarrow{\mathbf{E}}_{dist} = \{(\overrightarrow{\mathbf{r}} - \overrightarrow{\mathbf{v_i}}, \overrightarrow{\mathbf{v_j}} - \overrightarrow{\mathbf{r}}) : \text{ for each } (v_i, r, v_j) \in E\} \tag{6}$$

u_i^v and u_i^r are defined by the weighting scheme over the distributional model[1].

5   Distributional Navigation Algorithm

Once the KB is embedded into the τ-Space, the next step is to define the navigational process in this space that corresponds to a selective reasoning process in the KB. The navigational process is based on the semantic relatedness function defined as: $sr : VS^{dist} \times VS^{dist} \rightarrow [0, 1]$ is defined as:

$$sr(\overrightarrow{\mathbf{p_1}}, \overrightarrow{\mathbf{p_2}}) = \cos(\theta) = \overrightarrow{\mathbf{p_1}} . \overrightarrow{\mathbf{p_2}}$$

A threshold $\eta \in [0, 1]$ can be used to establish the desired semantic relatedness between two vectors: $sr(\overrightarrow{\mathbf{p_1}}, \overrightarrow{\mathbf{p_2}}) > \eta$.

The information provided by the semantic relatedness function sr is used to identify elements in the KB with a similar meaning from the reference corpus perspective. The threshold was calculated following the semantic differential approach proposed in [2]. Multiword phrases are handled by calculating the centroid between the concept vectors defined by each word.

Algorithm 1 is the Distributional Navigation Algorithm (DNA) which is used to find, given two semantically related terms $source$ and $target$ wrt a threshold η, all paths from $source$ to $target$, with length l, formed by concepts semantically related to $target$ wrt η.

The $source$ term is the first element in all paths (*line 1*). From the set of paths to be explored (*ExplorePaths*), the DNA selects a path (*line 5*) and expands it with all neighbors of the last term in the selected path that are semantically related wrt threshold η and that does not appear in that path (*line 7-8*). The stop condition is $sr(target, target) = 1$ (*line 10-11*) or when the maximum path length is reached.

The paths $p = <t_0, t_1, \cdots, t_l>$ (where $t_0 = source$ and $t_l = target$) found by DNA are ranked (*line 14*) according to the following formula:

$$rank(p) = \sum_{i=0}^{l} sr(\overrightarrow{\mathbf{t_i}}, \overrightarrow{\mathbf{target}}) \tag{7}$$

Algorithm 1 can be modified to use a heuristic that allows to expand only the paths for which the semantic relatedness between all the nodes in the path and the target term increases along the path. The differential in the semantic relatedness for two consecutive iterations is defined as $\Delta_{target}(t_1, t_2) = sr(\overrightarrow{\mathbf{t_2}}, \overrightarrow{\mathbf{target}}) - sr(\overrightarrow{\mathbf{t_1}}, \overrightarrow{\mathbf{target}})$, for terms t_1, t_2 and $target$. This heuristic is implemented by including an extra test in the line 7 condition, i.e., $\Delta_{target}(t_k, n) > 0$.

---

[1] Reflecting the word co-occurrence pattern in the reference corpus.

Algorithm 1. Distributional Navigation Algorithm

INPUT

- *threshold*: η
- *pair of terms* $(source, target)$ such that $sr(\overrightarrow{source}, \overrightarrow{target}) > \eta$
- *path length*: l

OUTPUT
$RankedPaths$: a set of ranked score paths $< (t_0, \cdots, t_l), score >$ such that $t_0 = source$ and $t_l = target$

```
 1: t_0 ← source
 2: Paths ← ∅
 3: ExplorePaths ← [(< t_0 >, sr(t⃗_0, target⃗))]
 4: while ExplorePaths ≠ ∅ do
 5:    remove (< t_0, ⋯, t_k >, sr(t⃗_k, target⃗)) from ExploredPaths
 6:    if k < l − 1 then
 7:       for all (n ∈ neighbors(t_k) : sr(n⃗, target⃗) > η and n ∉ {t_0, ⋯, t_k}) do
 8:          append (< t_0, ⋯, t_k, n >, sr(n⃗, target⃗)) to ExplorePaths
 9:       end for
10:    else if k = l − 1 then
11:       append (< t_0, ⋯, t_k, target >, 1) to Paths
12:    end if
13: end while
14: RankedPaths ← sort(Paths)
15: return RankedPaths
```

6   Evaluation

6.1   Setup

In order to evaluate the proposed approach, the τ-*Space* was built using the *Explicit Semantic Analysis* (ESA) as the distributional model. ESA is built over Wikipedia using the Wikipedia articles as *context co-occurrence windows* and TF/IDF as a weighting scheme.

ConceptNet[11] was selected as the commonsense knowledge base. *ConceptNet* is a semantic network represented as a labeled digraph $G_{ConceptNet}^{label}$ formed by a set of nodes representing concepts and a set of labeled edges representing relations between concepts. ConceptNet is built by using a combination of approaches, including open information extraction tools, crowd-sourced user input and open structured data. Concepts and relations are presented in the form of words or short natural language phrases. The bulk of the semantic network represents relations between predicate-level words or expressions. Different word senses are not differentiated. Two types of relations can be found: (i) recurrent relations based on a lightweight ontology used by ConceptNet (e.g. *partOf*) and (ii) natural language expressions entered by users and open information extraction tools. These characteristics make ConceptNet a heterogeneous commonsense knowledge base. For the experiment, all concepts and relations that were not in English terms were removed. The total number of triples used on the evaluation was 4,797,719. The distribution of the number of clauses per relation type is as follows: $= 1$ **(45,311)**, $1 < x < 10$ **(11,804)**, $10 \leq x < 20$ **(906)**, $20 \leq x < 500$ **(790)**, ≥ 500 **(50)**.

A test collection consisting of 45 (*source, target*) word pairs were manually selected using pairs of words which are semantically related under the context of the Question

Answering over Linked Data challenge (QALD 2011/2012)[2]. Each pair establishes a correspondence between question terms and dataset terms (e.g. 'What is the *highest* mountain?' where *highest* maps to the *elevation* predicate in the dataset). 51 pairs were generated in total.

For each word pair (a, b), the navigational algorithm 1 was used to find all paths with lengths 2, 3 and 4 above a fix threshold $\eta = 0.05$, taking a as source and b as target and vice-versa, accounting for a total of 102 word pairs. All experimental data is available online[3].

6.2   Reasoning Selectivity

The first set of experiments focuses on the measurement of the selectivity of the approach, i.e. the ability to select paths which are related and meaningful to the reasoning context. Table 1 shows the average *selectivity*, which is defined as the ratio between the *number of paths selected using the reasoning algorithm* 1 by the *total number of paths* for each path length. The total number of paths was determined by running a depth-first search (DFS) algorithm.

For the size of ConceptNet, paths with length 2 return an average of 5 paths per word pair. For this distance most of the returned paths tend to be strongly related to the word pairs and the selectivity ratio tend to be naturally lower. For paths with length 3 and 4 the algorithm showed a very high selectivity ratio (0.153 and 0.0192 respectively). The exponential decrease in the selectivity ratio shows the scalability of the algorithm with regard to selectivity. Table 1 shows the average selectivity for DNA. The variation of DNA with the Δ criteria, compared to DNA, provides a further selectivity improvement ($\phi = $ (# of spurious paths returned by DNA / # of spurious paths returned by DNA + Δ)) $\phi(length2) = 1$, $\phi(length3) = 0.49$, $\phi(length4) = 0.20$.

Table 1. Selectivity

Path Length	Average Selectivity Agorithm 1	% Pairs of Words Resolved	Path Acuracy
2	0,602	0,618	0,958
3	0,153	0,726	0,828
4	0,019	0,794	0,736

6.3   Semantic Relevance

The second set of experiments focuses on the determination of the *semantic relevance of the returned nodes*, which measures the expected property of the distributional semantic relatedness measure to serve as a heuristic measure for the selection of meaningful paths.

A gold standard was generated by two human annotators which determined the set of paths which are *meaningful* for the pairs of words using the following criteria: (i) all

---

[2] http://www.sc.cit-ec.uni-bielefeld.de/qald-1
[3] http://bit.ly/1p3PmHr

entities in the path are highly semantically related to both the source and target nodes and (ii) the entities are not very specific (unnecessary presence of instances, e.g. *new york*) or very generic (e.g. *place*) for a word-pair context. Only senses related to both source and target are considered meaningful.

The accuracy of the algorithm for different path lengths can be found in Table 1. The *high accuracy* reflects the effectiveness of the distributional semantic relatedness measure in the selection of meaningful paths. A systematic analysis of the returned paths shows that the decrease in the accuracy with the increase on path size can be explained by the higher probability on the inclusion of instances and classes with high abstraction levels in the paths.

From the paths classified as not related, 47% contained entities which are too specific, 15.5% too generic and 49.5% were unrelated under the specific reasoning context. This analysis provides the directions for future improvements of the approach (inclusion of filters based on specificity levels).

6.4   Addressing Information Incompleteness

This experiment measures the suitability of the distributional semantic relatedness measure to cope with KB incompleteness (gaps in the KB). 39 $<$ *source, target* $>$ entities which had paths with length 2 were selected from the original test collection. These pairs were submitted as queries over the ConceptNet KB indexed on the VS^{dist} and were ranked by the semantic relatedness measure. This process is different from the distributional navigational algorithm, which uses the relation constraint in the selection of the neighbouring entities. The distributional semantic search mechanism is equivalent to the computation of the semantic relatedness between the query (*source target*) and all entities (nodes) in the KB. The threshold criteria take the top 36 elements returned.

Two measures were collected. *Incompleteness precision* measures the quality of the entities returned by the semantic search over the KB and it is given by *incompleteness precision = # of strongly related entities / # of retrieved entities*. The determination of the *strongly related entities* was done using the same methodology described in the classification of the semantic relevance. In the evaluation, results which were not highly semantically related to both source and target and were too specific or too generic were considered incorrect results. The **avg. incompleteness precision value of 0.568** shows that the ESA-based distributional semantic search provides a feasible mechanism to cope with KB incompleteness, suggesting the discovery of highly related entities in the KB in the reasoning context. There is space for improvement by the specialization of the distributional model to support better word sense disambiguation and compositionality mechanisms.

The *incompleteness coefficient* provides an estimation of the incompleteness of the KB addressed by the distributional semantics approach and it is determined by *incompleteness coefficient = # of retrieved ConceptNet entities with an explicit association / # of strongly related retrieved entities*. The **average incompleteness value of 0.039** gives an indication of the level of incompleteness that commonsense KBs can have. The *avg. # of strongly related entities* returned per query is 19.21.

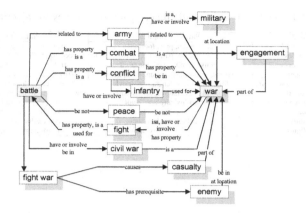

Fig. 2. Contextual (selected) paths between battle and war

An example of the set of new entities suggested by the distributional semantic relatedness for the pair $< mayor, city >$ are: **council, municipality, downtown, ward, incumbent, borough, reelected, metropolitan, city, elect, candidate, politician, democratic**.

The evaluation shows that distributional semantics can provide a principled mechanism to cope with KB incompleteness, returning highly related KB entities (and associated facts) which can be used in the reasoning process. The level of incompleteness of an example commonsense KB was analyzed and found to be high, confirming the relevance of this problem under the context of reasoning over commonsense KBs.

7   Analysis of the Algorithm Behavior

Figure 2 contains a subset of the paths returned from an execution of the algorithm for the word pair $< battle, war >$ merged into a graph. Intermediate nodes (words) and edges (higher level relations) provide a meaningful connection between the source and target nodes. Each path has an associated score which is the average of the semantic relatedness measures, which can serve as a ranking function to prioritize paths which are potentially more meaningful for a reasoning context. The output paths can be interpreted as an *abductive* process between the two words, providing a semantic justification under the structure of the relational graph. Table 2 shows examples of paths for lengths 2, 3 and 4. Nodes are connected through relations which were ommited.

The selectivity provided by the use of the distributional semantic relatedness measure as a node selection mechanism can be visualized in Figure 3 (A), where the distribution of the # of occurrences of the semantic relatedness values (y-axis) are shown in a logarithmic scale. The semantic relatedness values were collected during the navigation process for all comparisons performed during the execution of the experiment. The graph shows the discriminative efficiency of semantic relatedness, where just a tiny fraction of the entities in paths of length 2, 3, 4 are selected as semantically related to the target.

Table 2. Examples of semantically related paths returned by the algorithm

Paths - Length 2	Paths - Length 3	Paths - Length 4
daughter, parent, child	club, team, play, football	music, song, single, record, album
episode, show, series	chancellor, politician, parliament, government	soccer, football, ball, major_league, league
country, continent, europe	spouse, family, wed, married	author, write, story, fiction, book
mayor, politician, leader	actress, act_in_play , go_on_stage, actor	artist, create_art, work_of_art, art, paint
video_game, computer_game, software	film, cinema, watch_movie, movie	place, locality, localize, locate, location
long, measure, length	spouse, wife, marriage, husband	jew, religion, ethnic_group, ethnic, ethnicity
husband, married_man, spouse	aircraft, fly, airplane, pilot	war, gun, rifle, firearm, weapon
artist, draw, paint	country, capital, national_city, city	pilot, fly, airplane, plane, aircraft
city, capital, country	chancellor, head_of_state,	chancellor, member, cabinet,
jew, temple, religion	prime_minister, government	prime_minister, government

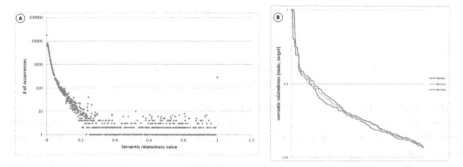

Fig. 3. # of occurrences for pairwise semantic relatedness values, computed by the navigational algorithm for the test collection (paths of length 2, 3, 4). Semantic relatedness values for nodes from distances 1, 2, 3 from the source: increasing semantic relatedness to the target.

In Figure 3(B) the average increase on the semantic relatedness value as the navigation algorithm approaches the target is another pattern which can be observed. This smooth increase can be interpreted as an indicator of a meaningful path, where semantic relatedness value can serve as a heuristic to indicate a meaningful approximation from the target word. This is aligned with the increased selectivity of the Δ (semantic relatedness differential) criteria.

In the DNA algorithm, the semantic relatedness was used as a heuristic in a greedy search. The worst-case time complexity of a DFS is $O(b^l)$, where b is the branching factor and l is the depth limit. In this kind of search, the amount of performance improvement depends on the quality of the heuristic. In Table 1 we showed that as the depth limit increases, the selectivity of DNA ensures that the number of paths does not increase in the same amount. This indicates that the distributional semantic relatedness can be an effective heuristic when applied to the selection meaningful paths to be used in a reasoning process.

8   Related Work

Speer et al. (2008) introduced AnalogySpace, a hybrid distributional-relational model over ConceptNet using Latent Semantic Indexing. Cohen et al.(2009) proposes PSI, a distributional model that encodes predications produced by the SemRep system. The τ-Space distributional-relational model is similar to AnalogySpace and PSI. Differences in relation to these works are: (i) the supporting distributional model (τ-Space is based on Explicit Semantic Analysis), (ii) the use of the reference corpus (the τ-Space distributional model uses an independent large scale text corpora to build the distributional space, while PSI builds the distributional model based on the indexed triples), (iii) the application scenario (the τ-Space is evaluated under an open domain scenario while PSI is evaluated on the biomedical domain), (iv) the focus on evaluating the selectivity and ability to cope with incompleteness. Cohen et al.(2012) extends the discussion on the PSI to search over triple predicate pathways in a database of predications extracted from the biomedical literature by the SemRep system. Taking the data as a reference corpus, Novacek et al.(2011) build a distributional model which uses a PMI-based measure over the triple corpora. The approach was evaluated using biomedical semantic web data.

Freitas et al.(2011) introduces the τ-Space under the context of schema-agnostic queries over semantic web data. This work expands the discussion on the existing abstraction of the τ-Space, defined in [1], introducing the notion of selective reasoning process over a τ-Space.

Other works have concentrated on the relaxation of constraints for querying large KBs. SPARQLer (Kochut et al. [10]) is a SPARQL extension which allows query and retrieval of semantic associations (complex relationships) in RDF. The SPARQLer approach is based on the concept of path queries where users can specify graph path patterns, using regular expressions for example. The pattern matching process has been implemented as a hybrid of a bidirectional breadth-first search (BFS) and a simulation of a deterministic finite state automaton (DFA) created for a given path expression. Kiefer et al.(2007) introduce iSPARQL, a similarity join extension to SPARQL, which uses user-specified similarity functions (Levehnstein, Jaccard and TF/IDF) for potential assignments during query answering. Kiefer et al.(2007) considers that the choice of a best performing similarity measure is context and data dependent. Comparatively the approach described on this work focuses a semantic matching using distributional knowledge embedded in large scale corpora while iSPARQL focuses on the application of string similarity and SPARQLer on the manual specification of path patterns.

9 Conclusion

This work introduced a selective reasoning mechanism based on a distributional-relational semantic model which can be applied to heterogeneous commonsense KBs. The approach focuses on addressing the following problems: (i) providing a semantic selection mechanism for facts which are relevant and meaningful in a specific querying and reasoning context and (ii) allowing coping with information incompleteness in large KBs. The approach was evaluated using ConceptNet as a commonsense KB and ESA as the distributional model and achieved *high selectivity, high selectivity scalability* and *high accuracy in the selection of meaningful paths*. Distributional semantics was used as a principled mechanism to cope with information incompleteness. An estimation of information incompleteness for a real commonsense KB was provided and the suitability of distributional semantics to cope with it was verified. Future work will concentrate on improving the accuracy of the proposed approach by refining the distributional semantic model for the selective reasoning problem.

Acknowledgments. This work has been joint funded by the Irish Research Council. This research was supported in part by funding from the Science Foundation Ireland under Grant Number SFI/12/RC/2289 (Insight). Joao C. P. da Silva is a CNPq Fellow - Science without Borders (Brazil).

References

1. Freitas, A., Curry, E., Oliveira, J.G., O'Riain, S.: Distributional Structured Semantic Space for Querying RDF Graph Data. International Journal of Semantic Computing 5(4), 433–462 (2011)
2. Freitas, A., Curry, E., O'Riain, S.: A Distributional Approach for Terminology-Level Semantic Search on the Linked Data Web. In: Proc. 27th ACM Symp. on Applied Computing (SAC 2012). ACM Press (2012)
3. Gabrilovich, E., Markovitch, S.: Computing semantic relatedness using Wikipedia-based explicit semantic analysis. In: Proc. of the 20th Intl. Joint Conf. on Artificial Intelligence, pp. 1606–1611 (2007)
4. Kiefer, C., Bernstein, A., Stocker, M.: The fundamentals of iSPARQL: A virtual triple approach for similarity-based semantic web tasks. In: Aberer, K., et al. (eds.) ASWC 2007 and ISWC 2007. LNCS, vol. 4825, pp. 295–309. Springer, Heidelberg (2007)
5. Turney, P.D., Pantel, P.: From frequency to meaning: vector space models of semantics. J. Artif. Int. Res. 37(1), 141–188 (2010)
6. Speer, R., Havasi, C., Lieberman, H.: AnalogySpace: Reducing the Dimensionality of Common Sense Knowledge. In: Proc. of the 23rd Intl. Conf. on Artificial Intelligence, pp. 548–553 (2008)
7. Cohen, T., Widdows, D., Schvaneveldt, R.W., Rindflesch, T.C.: Discovery at a Distance: Farther Journeys in Predication Space. In: BIBM Workshops, pp. 218–225 (2012)
8. Cohen, T., Schvaneveldt, R.W., Rindflesch, T.C.: Predication-based Semantic Indexing: Permutations as a Means to Encode Predications in Semantic Space. In: T. AMIA Annu. Symp. Proc., pp. 114–118 (2009)
9. Novacek, V., Handschuh, S., Decker, S.: Getting the Meaning Right: A Complementary Distributional Layer for the Web Semantics. In: Proc. of the Intl. Semantic Web Conference, pp. 504–519 (2011)
10. Kochut, K.J., Janik, M.: SPARQLeR: Extended SPARQL for semantic association discovery. In: Franconi, E., Kifer, M., May, W. (eds.) ESWC 2007. LNCS, vol. 4519, pp. 145–159. Springer, Heidelberg (2007)
11. Liu, H., Singh, P.: ConceptNet A Practical Commonsense Reasoning Tool-Kit. BT Technology Journal 22(4), 211–226 (2004)
12. Harris, Z.: Distributional structure. Word 10(23), 146–162 (1954)
13. Speer, R., Havasi, C., Lieberman, H.: AnalogySpace: Reducing the Dimensionality of Common Sense Knowledge. In: Proc. of the 23rd Intl. Conf. on Artificial Intelligence, pp. 548–553 (2008)

Ontology Translation: A Case Study on Translating the Gene Ontology from English to German

Negacy D. Hailu, K. Bretonnel Cohen, and Lawrence E. Hunter

Computational Bioscience Program, University of Colorado School of Medicine, USA
{negacy.hailu,larry.hunter}@ucdenver.edu, kevin.cohen@gmail.com

Abstract. For many researchers, the purpose of ontologies is sharing data. This sharing is facilitated when ontologies are available in multiple languages, but inhibited when an ontology is only available in a single language. Ontologies should be accessible to people in multiple languages, since multilingualism is inevitable in any scientific work. Due to resource scarcity, most ontologies of the biomedical domain are available only in English at present. We present techniques to translate Gene Ontology terms from English to German using DBPedia, the Google Translate API for isolated terms, and the Google Translate API for terms in sentential context. Average fluency scores for the three methods were 4.0, 4.4, and 4.5, respectively. Average adequacy scores were 4.0, 4.9, and 4.9.

1 Introduction

For many researchers, the purpose of ontologies is sharing data. This sharing is facilitated when ontologies are available in multiple languages, but inhibited when an ontology is only available in a single language. There are public health consequences for the lack of availability of ontologies in multiple languages, as well—Rebholz-Schuhmann has posited that multilinguality will be the the key to transferring the results of genomic research to actual patient care (personal communication). There would thus be considerable significance to the development of the ability to translate biomedical ontologies from one language to another. However, manual translation of biomedical ontologies is unlikely to scale, for a number of reasons. Translation of biomedical linguistic resources may require considerable domain expertise in addition to strong language skills. Biomedical ontologies change frequently, making manual updating difficult. For these reasons, the feasibility of machine translation of biomedical ontologies is an important research question.

A number of hypotheses are evaluated in the work reported here. The most basic hypothesis is that machine translation of biomedical ontologies at an acceptable level of fluency and adequacy is possible. A further hypothesis is that the problems of ambiguity, domain specificity, and lack of context of terms that come from an ontology can be overcome by providing appropriate linguistic context. The final hypothesis is that context can be provided automatically by retrieval of sentences from a domain-specific document collection.

E. Métais, M. Roche, and M. Teisseire (Eds.): NLDB 2014, LNCS 8455, pp. 33–38, 2014.
© Springer International Publishing Switzerland 2014

German was chosen as the target language for the machine translation effort. There were two motivations for this. The first is that German is the second-most represented language in PubMed/MEDLINE, after English [14]. Thus, there are many opportunities for mapping English-language scientific publications to German-language publications if ontology terms can be translated. The second is that Germany is a Tier 1 market for electronic medical records, with a projected hospital-based electronic medical record adoption rate of 35% in 2013 [8]. Thus, there is a large opportunity for penetrance of basic research into the German health care system if the ontology translation problem can be solved.

As input for the machine translation task, the Gene Ontology [9] was selected. The Gene Ontology is the flagship ontology in the biomedical domain. It contains about 38,000 concepts describing the broad categories of molecular function, biological process, and cellular component. It is essentially the lingua franca for comparing gene functions across and within organisms. It is used by many large biological database efforts for the annotation and classification of genes in multiple organisms. Gene Ontology annotations (mappings between genes and Gene Ontology concepts) are maintained and distributed by some of the largest genomic databases in the world—millions of them are available through the Gene Ontology Annotation [9] database.

2   Research Strategy

To understand the nature of the data, two foundational questions were investigated: the distribution of GO terms with respect to word length, and the coverage of GO terms in DBPedia.

Both the full Gene Ontology and the species-neutral GO slim were sampled for evaluation. GO "slims" are subsets of GO terms that are cleaner and give a broader overview of the ontology content, as opposed to the full Gene Ontology.

This was motivated to test the naive hypothesis that if a GO term is single word, then we can translate it using a dictionary—the two most popular English to German online dictionaries that we used are www.dict.cc and www.leo.org. However, for terms with multiple words, we hypothesized that Wikipedia and Google Translate API can be used to do the translation.

The null hypothesis that there is no difference within the two group of terms in frequency in terms of word length was tested using the non-parametric *Wilcoxon Rank Sum Test*. At a $p - value < 0.5$, we failed to reject the null hypothesis. Since the statistical test showed us there is no difference between the GO terms and GO slims with respect to word length, we focused testing and evaluation on terms sampled from the GO slims.

The fact that single word terms could be translated using dictionaries sounds very intuitive but only 13% of the GO slims are a single word. We remained with the large portion that cannot be translated using dictionaries. There are also other issues that remained critical while using dictionaries to translate single word terms. The main issue is *domain specificity*—for example, www.dict.cc gives us 15 translations for the word *cell*—being the second translation for *cell phone*. We concluded that dictionaries don't give us domain specific translation. In section 3, we discuss how we use Wikipedia and Google Translate API to translate multiple word terms.

3   Methods

Machine translation is traditionally done in two ways—Statistical Machine Translation (SMT) and Rule-based Machine Translation (RBMT). The strength of RBMT is that it doesn't require huge computational resources and as stated in [1] they are good for languages with limited resources. However, they are based on linguistic theories and it is costly to get human experts in the field. Also, inconsistency among the experts is a big challenge. The strength of SMT is that it works well and can be done without domain or linguistic expertise. However, it requires large collections of parallel texts. These two most popular techniques won't work for our case for two reasons—first of all we don't have human experts who can design linguistic rules to translate ontologies. Secondly, ontologies are mostly noun phrases and it doesn't make sense to have a parallel bilingual corpus of single or short noun phrases to train them using SMT.

We used DBpedia—structured Wikipedia [2] to translate GO terms from English to German. We made keyword based search for each term in the DBpedia dataset. If a term has both English and German Wikipage, we pick the title of that term in both pages as a translation. It turns out that 25% of the terms have a wiki page in both languages. The fact that the Wikipedia URI look up was keyword based made a small portion of the terms to appear with their correspondent Wikipedia page.

Studies show that Gooogle Translate API is widely used by many users in day to day life usually on a sentence level translation though its accuracy widely vary from language to language [3]. Although it is very popular for general-domain translation, Steedman has shown that it is easy to reveal systematic sources of problems with it even in the general domain [4]. We were curious how this translation tool will perform in translating terms in certain domain. Google Translate API is a statistical machine translation model which is trained on huge corpus collected from different sources but mainly from online news. It was not primarily trained on biomedical literature. We hypothesize that Google Translate API gives better translation for full sentences than for short terms in certain domain. In order to test this hypothesis we need to prepare two dataset—GO slims in their native form, and within a sentence. We searched PubMed/Medline by full text match to collect sentences that have GO slims in them. We collected more than 1000 sentences in this way. We translated both the GO slims in their native form and in sentences to German using Google Translate API.

4   Results

Evaluation was done by human experts in the field and who are fluent in both languages. We prepared a randomized dataset for evaluation as shown in Table 1a in the three format—Wiki translation, Google Translate API without context i.e. terms in themselves, Google Translate API with context i.e. terms within a sentential form. The evaluators were asked to evaluate each translated term for *adequacy* and *fluency* in a likert scale of 1 through 5 where 1 is poor and 5 is excellent. Adequacy addresses how much information is translated between the original and translated text? Fluency will tell us how good is the German sentence/phrase? Our human based evaluation for the three methods is shown in

Table 1.

(a) Evaluated dataset size

Methods	Total terms	Evaluated terms
Wikipedia translation	75	75
Google Translate API without context	1000	100
Google Translate API with context	1000	100
Total terms	2075	2075

(b) Average likert scale score

Methods	Group-1		Group-2	
	Adequacy	Fluency	Adequacy	Fluency
Wiki translation	4.3	4.3	3.9	3.7
Google Translate API without context	4.9	4.6	4.8	4.2
Google Translate API with context	4.9	4.5	4.9	4.4

Table 1. Two groups each having two people participated in the evaluation. The scores reported by both groups for the last two methods i.e. Google Translate API with and without context is pretty similar. However, the evaluators reported that Wiki translation performing relatively low—part of the reason is the way we used DBPedia as a translation. We followed keyword based exact match query on the DBPedia dataset. It is well known that such approach will not give you hit for example with acronyms like ADPG pyrophosphorylase complex.

The goal of the third method—constructing sentences that include gene ontology terms and using Google Translate API to translate them was meant to improve semantics in translation. From the evaluation result in Table 1, there appears to be no difference on the results for the two methods. Even though previous works using Statistical Machine Translation to translate ontologies [5] and [6] concluded context plays a big role in ontology translation, we found out that there is no difference at all.

Terms like *bleb*—an irregular bulge in the plasma membrane of a cell, was not properly translated by any of the methods. However, it was correctly translated as *zeiose* with the help of its English synonym *zeiosis*. This shows that synonyms could play a big role in translation.

It is very important to discuss if the relationships between concepts in an ontology remain constant when the terms in the ontology are translated into arbitrary languages? In theory, they should, since the concepts are specified by their definitions, not by the associated terms. However, whether or not this is the case has not been evaluated. The availability of complete translated ontologies produced by methods like the one described here would enable investigation of this question.

Furthermore, we questioned why longer terms are translated better than shorter terms even when no additional context is provided? We speculate that the longer terms themselves provide "intrinsic" context due to their larger amount of material, thus allowing better disambiguation, etc. is possible for shorter terms in isolation.

We noted large differences in performance between DBPedia and the Google Translate API, but not between the performance of the Google Translate API with and without additional context. There is neither an improvement nor a degradation of performance when additional context is provided. From this we conclude that the extra work of providing context may not be required for translating biomedical ontologies, although further work is required to see if this result holds for other ontologies.

5   Conclusion

Ontology terms can be a single word or as long as a complex phrase. Translating the short terms—terms which have word length less than or equal to 3—is difficult, since they are ambiguous. In this work, we constructed sentences using the ontologies so that the terms would have some contextual cues as to word sense. The sentence construction was done using the terms as key words in a PubMed/MEDLINE search. Evaluation was done by human experts. The evaluators were given a Likert scale ranging from 1 through 5, where 1 is poorly translated and 5 is fluent. They were asked to assess adequacy and fluency. On average, the adequacy Likert scale scores were 4.0, 4.9, and 4.9 for DBPedia, GTA without context, and GTA with context. Average fluency scores for the three methods were 4.0, 4.4, and 4.5. Our results showed that performance for translation of short terms was not improved when the terms are in a sentential context rather than in isolation.

The work reported here describes experiments and analyses that add to our understanding of the challenges and opportunities for automatic translation of biomedical ontologies. Overall, the results are quite encouraging, with high adequacy and fluency achievable for the terms associated with most concepts. Automatic translation overcomes the issues of scale and updating in biomedical ontologies, and enables research into the fundamental issues concerned with translated ontologies, such as preservation of structure and relationships in the second language.

Acknowledgments. The authors thank Carsten Goerg, Barbara Grimpe, Daniel Pape, Christophe Roeder, Kai Stuehler, and Daniel Waldera-Lupa for their work on the evaluation.

References

1. Costa-Jussà, M.R., Farrús, M., Mariño, J.B., Fonollosa, J.A.R.: Study and Comparison of Rule-based and Statistical Catalan-Spanish Machine Translation Systems. Computing and Informatics 31, 245–270 (2012)

2. Lehmann, J., Isele, R., Jakob, M., Jentzsch, A., Kontokostas, D., Mendes, P.N., Hellmann, S., Morsey, M., van Kleef, P., Auer, S., Bizer, C.: DBpedia A Large-scale, Multilingual Knowledge Base Extracted from Wikipedia. In: Semantic Web (2012)
3. Aiken, M., Balan, S.: An Analysis of Google Translate Accuracy. Translation Journal 16(2) (2013)
4. Steedman, M.: On becoming a discipline. Computational Linguistics 34(1), 137–144 (2008)
5. Arcan, M., Buitelaar, P.: Ontology Label Translation. In: HLT-NAACL (2013)
6. McCrae, J., Espinoza, M.: Combining statistical and semantic approaches to the translation of ontologies and taxonomies. In: Fifth Workshop on Syntax, Semantics and Structure in Statistical Translation, ACL HLT 2011, pp. 40–46. The Association for Computational Linguistics (2013)
7. Brown, P.F., Pietra, V.J.D., Pietra, S.A.D., Mercer, R.L.: The mathematics of statistical machine translation: parameter estimation. Computational Linguistics 19(2) (June 1993)
8. Accenture, Overview of international EMR/EHR markets: Results from a survey of leading health care companies (2010)
9. Ashburner, M., Ball, C., Blake, J., Botstein, D., Butler, H., Cherry, J.M., Davis, A., et al.: Gene ontology: tool for the unification of biology. The Gene Ontology Consortium, Nature Genetics 25(1), 25–29 (2000)
10. Aggarwal, N., Asooja, K., Buitelaar, P.: DERI&UPM: Pushing corpus based relatedness to similarity: Shared task system description. In: SemEval 2012 (2012)
11. Denkowski, M., Lavie, A.: Meteor 1.3: Automatic Metric for Reliable Optimization and Evaluation of Machine Translation Systems. In: Proceedings of the Sixth Workshop on Statistical Machine Translation, pp. 85–91. Association for Computational Linguistics, Edinburgh (2011)
12. Gabrilovich, E., Markovitch, S.: Computing semantic relatedness using Wikipedia-based explicit semantic analysis. In: In Proceedings of The Twentieth International Joint Conference for Artificial Intelligence, Hyderabad, India, pp. 1606–1611 (2007)
13. Papineni, K., Roukos, S., Ward, T., Zhu, W.-J.: BLEU: A Method for Automatic Evaluation of Machine Translation. In: Proceedings of the 40th Annual Meeting of the Association for Computational Linguistics, ACL 2002, pp. 311–318. Association for Computational Linguistics, Stroudsburg (2002)
14. Jimeno-Yepes, A., Prieur-Gaston, É., Névéol, A.: Combining MEDLINE and publisher data to create parallel corpora for the automatic translation of biomedical text. BMC Bioinformatics 14(1), 146 (2013)
15. Doddington, G.: Automatic evaluation of machine translation quality using n-gram co-occurrence statistics. In: Proceedings of the Second International Conference on Human Language Technology Research, HLT 2002, pp. 138–145 (2002)

Crowdsourcing Word-Color Associations

Mathieu Lafourcade[1], Nathalie Le Brun[2], and Virginie Zampa[3]

[1] LIRMM, Montpellier, France
[2] Imagin@t, Lunel, France
[3] LIDILEM, Grenoble, France

Abstract. In Natural Language Processing and semantic analysis in particular, color information may be important for processing textual information. Knowing what colors are generally associated with terms by people is valuable. We explore how crowdsourcing through a game with a purpose (GWAP) can be an adequate strategy to collect such lexico-semantic data.

Keywords: Word Color Associations, Lexical Network, Crowdsourcing.

1 Introduction

Color information is important in our daily life and may be of interest in the context of Natural Language Processing. Although this is beyond the scope of this paper, they are strong connections between colors and emotions. However, provide information about word-colors associations to a system dedicated to the semantic analysis of texts, in addition to other traditional knowledge (hypernym, parts of, semantic role, etc.) could greatly improve system performance. Association between word and color could be made for abstract nouns related to emotions (like fear, anger, danger, hope, ...) but in a more straightforward way for concrete nouns (like sky, lion, snow, sea, ...)

There is a very lively debate as to whether the associations between color and meaning were independent of age, gender, or nationality. This might be the case, for example, for the relation between *red* and *danger*, since the red may be related to *blood/fire* regardless of others factors. Berlin and Kay (1969) argued that differences could be organized into a coherent hierarchy, and that there is a limited number of *universal* basic color terms used in various cultures. Berlin and Kay based their analysis on a comparison of color words in 20 languages from around the world, but their findings have been discussed a lot. In the same way, several expressions using colors have the same meaning in different languages especially when they are culturally close, which is hardly a surprise. For example, *dark thoughts* in English and *idées noires* are roughly equivalent, as well as *to see red* and *voir rouge*. Many studies, mostly in English, have been undertaken to determine relations between colors and words or colors and emotions, etc. Most of those studies of psycholinguistic are undertaken in a classical way, and their raw results are general modest in size and not freely available. Furthermore, as previously mentioned, it is extremely delicate (and probably unwise) to translate directly the result of such studies from one language to another. Finally, we can say that there is no definite consensus on the universality of word-color associations for abstract or feeling words.

E. Métais, M. Roche, and M. Teisseire (Eds.): NLDB 2014, LNCS 8455, pp. 39–44, 2014.

Mohammad (2011b) conducted experiments using eleven colors and showed that more 30% of the terms have a strong color association (for a lexicon of everyday words, not including specialized domains). About 33% of thesaurus categories (like Roget) have color associations, and abstract terms are associated with colors almost as often as physical entities do, mostly by metaphor. Again, Nijdam (2010) compares different models on relation between color and emotion and proposes a correspondence between emotions and colors. He concludes that some models about color-meaning may have some overlap but they also show a great amount of vagueness, certainly because the color is dependant of personal/cultural situation. In summary, the acquisition of data on associations between words and colors is hampered by high variability, (even restricted to a given language). The more abstract is the term, the higher the variation is to be anticipated. Gözde et al. (2011) make a resource that contains information about the association of words and colors in English. They made a short selection of 200 words, a subset of words used in Grefenstette (2005), and compare annotations made with Amazon Mechanical Turk service (10 annotators – 11 colors) with three automatic methods: image analysis, language models on web data and similarity between words and colors (using LSA).

At the present time, as far as we know, Gözde et al. (2011) are the only ones who tried to make a resource about association of colors and words in English. Such resources are very rare in other languages, and especially it doesn't exist in French language. But as noted by Grefenstette (2005) what people generally know about things (concrete things, but even abstract concepts) includes their typical color as an important component. That is why, this information is not usually shown in dictionary definitions or lexical resources, although it would be useful for various computer applications, and in NLP in particular. We think that color information can be very helpful in the context of automated lexical disambiguation (Word Sense Disambiguation - WSD). For example, the French word *tissu* is polysemous and can mean either *fabric* or *living tissues*. If in a given text, *tissu* is associated to *bleu* (blue), the color information can help to choose the right meaning and at least eliminate the wrong ones. Another computer application for which color information would be helpful is one that allows you to find a momentarily forgotten word but "on the Tip of the Tongue" (Lafourcade 2012 and Joubert 2012). The color, either actual or symbolic, is very often crucial for finding proper terms.

The objective of this paper is to present how to easily and efficiently produce a (French) resource that lists associations between colors and words, without resorting to corpora. For this purpose, we implemented a GWAP (game with a purpose as dubbed by L. von Ahn. (2006)) named ColorIt through which the player is asked to tell the color he spontaneously associates with a given term (http://www.jeuxdemots. org/colorit.php). At first we show how information on color is a valuable asset in the context of NLP, and then we detail the features of the game as well as some quantitative and qualitative results.

2   ColorIt, a Game for Collecting Colors

The goal of ColorIt is to collect spontaneous associations between colors and words, colors assigned to concrete terms as well as those symbolically and subjectively associated to verbs or nouns denoting abstract entities.

A word is presented to the user along with a choice of colors. The user is invited to click on the color he associates to the displayed word. It is possible to associate *no color* (if color is not applicable) or to pass (if the player does not know the word for instance). Passing is not penalized. As an alternative to the color palette, a text field is (optionally) proposed in which experienced player can enter *several* colors (as free text). Once the choice is done, a notification indicates the score along with the answers of other players for the same word. The score depends on the adequacy of the response of the player and the color distributions already assigned to the word *via* other answers. There are two types of answers: specify one or several colors or choose *no color/not applicable*. If the player's choice corresponds to the other votes, then he gets some points (max 50), otherwise he loses some points (max 100). The amount of points is related to the distribution of colors related to the word in the lexical network. If a color is mildly associated (compared to others), the player earns some points. If the color is very weakly associated, the player does not gain many points. The associated color weights are increased by one for each vote. A newly introduced color is not rewarded (0 vote before), but will be rewarded afterward; it enters the lexical network with a weight of 1. For example (very simplified considering the current distribution), suppose the term *elephant* has the following colors with weights: *gray/24, pink/10, white/5*. If the player selects *grey* he will win 24 points and *grey* is set to 25. If the player selects *no color* he loses 39, but the *no color* relation is introduced in the lexical network with a weight of 1 for this term.

ColorIt is a quite challenging game, and we received very positive feedback from players. Beside some terms whose color is obvious (*snow, night, coal,* etc.) and no lend itself to interpretation, some others may have a wide range of possible colors either objective for concrete nouns (like *flower, car,* etc.) or subjective for abstract concepts (such as *anger, sadness,* etc.) or no color at all (*increase, subjectivity, forwardness,* etc.). Choose a color that is believed to be the most relevant is not an easy task and the confrontation with the other answers may generate excitement, suspense, and surprise ...either good or bad ! Although the purpose of this game is to collect lexico-semantic data, no specific linguistic knowledge is required to play, and the data collected are of good quality.

The JDM project (Lafourcade, 2007 and Chamberlain, 2013) aims at building a very large lexical and semantic network through games (GWAP, Games With A Purpose) and crowdsourcing. Such a network is composed of lexical items as nodes (vertices), and relations between nodes. Relations are typed, oriented and weighted. The lexical network for French has been made freely available to the community, so as to be an interesting resource to work with. The resource contains over 300 000 terms and 5 millions relations. The JDM lexical network contains terms, word meanings and various semantic information (like person, living being, abstract, concrete, artefact, ...). Relations are weighted (the most obvious or frequent having the higher value).Some relations may have a negative weight to represent an interesting impossibility or exception (like for example *ostrish --can:-100-- fly*). Some impossible colors can be represented with such an approach. The word meanings are connected to the main word with a refinement relation. For example; *frégate* (frigate) is connected to *frégate>navire* (frigate ship) and *frégate>oiseau* (frigate bird). The term *frégate>navire* is itself refined as *frégate>navire>moderne* (modern frigate ship) and *frégate>navire>ancien* (ancient frigate ship). Each refinement may itself be connected to several other terms in the network. In this network, the information about color existed until now only through the *characteristics* relation, thus mixed with other information. Now it is the subject of a specific relation that the game ColorIt enriches consistently. A heuristic algorithm was designed to have a reasonable probability of finding a term for which the color information is meaningful. Indeed, we avoid to propose too many terms for which the color information would be irrelevant, so as not to deter players.

The words which are displayed to the player are selected *via* the algorithm as follows: we select randomly, in the *JeuxDeMots* lexical network, a word that has at least one color relation, and we propose, through a virtual coin tossing, either the term or one of its linked neighbors in the network. This causes a rapid spread of color information across the network. A term, that has been repeatedly tagged as *without color* will be removed from the list of selectable neighbors. In the same way, the refinements of "colored" words quickly become "colored" themselves. The assumption behind this *propagation algorithm* is that there is more chance to select a term eligible for color information if it is linked to a term having already a color characteristic. A completely random selection would be counterproductive because most of the time a color characteristic would not be applicable. In the JDM lexical network, we estimate by sampling that only about 10% of the approximately 300.000 terms (30,000 out of over 300,000) are "color eligible".

If a response provided by the player is in the lexical network but is not known as a color/appearance, it is then proposed and will be validated (or invalidated) as such by an administrator. But if the proposed term does not exist at all in the network, it must first be integrated in the network before being characterized as a color or appearance (again by an administrator). The Diko interface (www.jeuxdemots.org/diko.php) of JDM is a contributory tool for validation of proposals. The propagation algorithm has been bootstrapped with the colors proposed in the game: *white, grey, white, red, dark red, orange, yellow, light green, green, blue, light blue, purple, pink, beige, brown.* Before the creation of ColorIt, these colors were associated with terms through other relations than color.

3   Data Collected and Evaluation

After 3 months (from August to November 2013), more than 32,000 word-color(s) associations have been created and more than 15,000 eligible terms were provided with color information. So, on average, one term is associated with 2.2 colors. The number of votes exceeded 300,000. We estimate that half of the terms in the JDM lexical network that are color eligible (15000 out of over 30.000) have been "colorized" As soon as a color is associated to a term, the network is updated with this color or appearance. For example, *eau* (water) is associated with different blues (*light blue* (26), *blue* (9), *blue lagoon* (1), *turquoise* (1), etc.) many colors (*greenish* (4), *green* (2), etc.) and different appearances (*cloudy* (20), *transparent* (9+9), *crystal clear* (4), *colorless* (4), etc.). With the possibility to add a new color or appearance *via* the free text field, the network has been completed with names of specific colors like *yellow carthusian, tangerine,* etc. or appearances like *striped, translucent,* etc. or with new combination of colors. At the present time, more than 700 terms refers to colors or appearance.

The data collected were evaluated by hand, on two random samples of around 500 terms, totalizing around 2,500 term-color associations. The first sample S_C was taken from the terms having at least one color associated the second sample S_{NC} amongst those with at least one vote *no color*. Terms in S_C may have some *no color* associations and conversely, terms in S_{NC} might have some colors associations. We asked volunteers to examine randomly some word-color associations and tell whether they were correct or not (either as color or *no color*). The evaluation approach, where people are asked to judge word-color associations (closed task) is an opposite process compared to the acquisition where people are asked to produce associations (open task). However, this evaluation protocol does not inform on the color combinations that may have been omitted.

Table 1. Percentages of correctly and incorrectly assigned colors

	Proper colors	Wrong colors	Proper no color	Wrong no color	Total
# votes	13308	366	708	81	14463
%	92 %	2,5 %	5 %	0,5 %	100

The evaluation by the volunteers covered about 10% of term-color associations generated by the players (acquisition).The collected data seem to contain a quite low percentage of wrong associations (see Table 1). Furthermore, closer inspection of the lexical network shows that wrong associations are assigned a low weight compared to the correct associations (less votes). For example, for 1 vote as *blue* for *sun*, we have more than 100 votes as *yellow*.

Table 2. Kappa between players according to semantic field (class) to which the word belongs. In parentheses: the number of terms related to this semantic field in the lexical network.

Term related to	Politic (2702 terms)	Zoology (3191 terms)	Arts (2815 terms)	Emotions (304 terms)	Plants (4481 terms)
agreement	0.55	0.65	0.71	0.28	0.76

Table 2 presents the evaluation of the agreement between players based on some topics (semantic fields) of the suggested words. The agreement value for a given term is the mean of the Cohen kappa between players associations taken in pairs for this term. The agreement value for a class is the mean of the agreements of the terms of this class (those that have been played). Some terms may relate to more than one class. This is for the semantic field related to plants that the agreement between the players is the highest. It appears that many players refer to external sources (like Wikipedia) to find out what color(s) must be assigned to an exotic or unknown plant or animal. The weakest agreement is that related to the semantic field of emotions, which is easily explained by the high degree of subjectivity of corresponding associations (and this although our experiment has been conducted only in French, and with a relatively homogenous population consisting of 70% women between 30 and 50). On the semantic field of the arts, there are many color names expressed through specific vocabulary, but once decrypted finally have quite a few variations.

4   Conclusion

We have designed a simple but challenging and effective game for collecting data about word-color associations. In a few months more than 15 000 terms were associated with one or more colors, and about 4,000 were tagged as irrelevant for color information. Many colors and appearances have been introduced in the

JeuxDeMots lexical network *via* the game ColorIt. Currently, as far as we know, the data collected are the first French language resource for this type of associations. Furthermore, this resource is freely available dynamically from the website. One of the challenges in the game design was to achieve a propagation algorithm allowing the selection of eligible terms for color information, so as to avoid boring players in offering too many terms to which it is not appropriate to assign a color. We had several surprises with the data collected with ColorIt. First, choose "no color" was originally designed to be a loophole for players when nothing suitable. But it soon became clear that the feature "colorless" was precious to eliminate certain meanings when polysemy. Indeed, in the context of word sense disambiguation, selection of meaning by the indications of color is more often done by elimination than by selection. One can explain this, at least partially, by the polysemy metaphorically: the abstract sense (without color) often comes from the concrete term. Secondly, a side effect of the project was to add a large number of color names and appearances that were beforehand unknown in the lexical network In addition to newly introduced true colors, the network now contains numerous terms related to the visual aspect, like *translucent*, *stripped*, *blurred*, *scratched*, *spotted*, etc. To go further, another perspective might be to evaluate the correlation between color and feelings associated with certain words, especially abstract nouns. Once we have a sufficient number of words-colors associations, the comparison with other resources already available on the polarity, or feelings / emotions associated might be conducted without much difficulty, without forgetting their use in the context of word sense disambiguation.

References

1. von Ahn, L.: Games with a purpose. Computer 39(6), 92–94 (2006)
2. Berlin, B., Kay, P.: Basic Color Terms: Their Universality and Evolution. University of California Press, Berkeley (1969)
3. Chamberlain, J., Fort, K., Kruschwitz, U., Lafourcade, M., Poesio, M.: Using Games to Create Language Resources: Successes and Limitations of the Approach. In: Gurevych, I., Kim, J. (eds.) Theory and Applications of Natural Language Processing, p. 42. Springer (2013) ISBN 978-3-642-35084-9, 2013
4. Grefenstette, G.: The color of things: Towards the automatic acquisition of information for a descriptive dictionary. Revue Française de Linguistique Applique 2, 83–94 (2005)
5. Joubert, A, Lafourcade, M.: A new dynamic approach for lexical networks evaluation. In: Proc. of the Eight International Conference on Language Resources and Evaluation (LREC 2012), Istanbul, Turkey, May 23-25 (2012)
6. Lafourcade, M.: Making people play for Lexical Acquisition. In: Proc. SNLP 2007, 7th Symposium on Natural Language Processing, Pattaya, Thailande, December 13-15, p. 8 (2007)
7. Lafourcade, M., Joubert, A.: Increasing long tail in weighted lexical networks. In: Proc. of Cognitive Aspects of the Lexicon (CogAlex-III), Coling, Mumbai, India (December 2012)
8. Meier, B.: Ace: a color expert system for user interface design. In: Proceedings of the 1st Annual ACM SIGGRAPH Symposium on User Interface Software, UIST 1988, pp. 117–128. ACM, New York (1988)
9. Nijdam, N.A.: Mapping Emotion to Color, Human Media Interaction Human Media Interaction. University of Twente, The Netherlands (2010),
   http://hmi.ewi.utwente.nl/verslagen/capitaselecta/
   CSNijdamNiels.pdf

On the Semantic Representation and Extraction of Complex Category Descriptors

André Freitas[1], Rafael Vieira[2], Edward Curry[1], Danilo Carvalho[3],
and João Carlos Pereira da Silva[2]

[1] Insight Centre for Data Analytics, National University of Ireland, Galway
[2] Computer Science Department, Federal University of Rio de Janeiro (UFRJ)
[3] PESC/COPPE, Federal University of Rio de Janeiro (UFRJ)

Abstract. Natural language descriptors used for categorizations are present from folksonomies to ontologies. While some descriptors are composed of simple expressions, other descriptors have complex compositional patterns (e.g. 'French Senators Of The Second Empire', 'Churches Destroyed In The Great Fire Of London And Not Rebuilt'). As conceptual models get more complex and decentralized, more content is transferred to unstructured natural language descriptors, increasing the terminological variation, reducing the conceptual integration and the structure level of the model. This work describes a representation for complex natural language category descriptors (NLCDs). In the representation, complex categories are decomposed into a graph of primitive concepts, supporting their interlinking and semantic interpretation. A category extractor is built and the quality of its extraction under the proposed representation model is evaluated.

1 Introduction

Ontologies, vocabularies, taxonomies and folksonomies provide structured descriptors for categories of objects and their relationships. While ontologies target a more centralised, consistent and structured representation of a domain, folksonomies allow a decentralised, less structured categorization. Both representation models have in common natural language descriptions associated with object categories. These *natural language category descriptors* (NLCDs) are a fundamental part of the communication of the meaning behind these artefacts. While some descriptors are composed of single words or simple expressions (e.g. 'Person', 'Country', 'Film'), other descriptors have more complex compositional patterns (e.g. 'French Senators Of The Second Empire', 'United Kingdom Parliamentary Constituencies Represented By A Sitting Prime Minister').

As models get more complex and decentralized, more content is transferred to unstructured natural language descriptors, increasing the terminological variation, reducing the conceptual integration and the structure level of the model. In this scenario, the more formal conceptual model tools are substituted by complex NLCDs as an interface for domain description. From the perspective of information extraction and representation, NLCDs provide a much more tractable

E. Métais, M. Roche, and M. Teisseire (Eds.): NLDB 2014, LNCS 8455, pp. 45–50, 2014.

subset of natural language which can be used as an '*interface*' for the creation of structured domains. From the syntactic perspective, natural language category descriptors (NLCDs) are short and syntactically well-formed phrases. Differently from full sentences, NLCDs present simpler and more regular compositional patterns. By structuring NLCDs, we intend to support the creation of more structured resources with lower construction effort and in a more decentralized way.

In this work we describe a representation and an extraction approach for complex NLCDs. In the representation, complex predicates are decomposed into a graph of primitive word senses supporting the alignment between different NLCDs. A NLCD extractor is built and the extraction quality is evaluated. An extended version of this paper can be found at the website[1]

2   Related Work

Different works have focused on information and data extraction approaches applied in the context of semantic annotations and the Semantic Web. Most of these approaches have targeted the extraction of ontologies and datasets from semi-structured data [1], from unstructured data [7] or the alignment of folksonomies to ontologies [3][4][6]. YAGO [1] is a large-scale ontology which is automatically built from Wikipedia and WordNet. YAGO extracts facts from the infoboxes and the category system of Wikipedia, representing them in a data model which is based on reified RDF triples. YAGO builds a taxonomic structure from Wikipedia categories, aligning them to WordNet synsets. Specia & Motta [3] propose an approach for making explicit the semantics behind the tags, by using a combination of shallow pre-processing strategies and statistical techniques, together with knowledge provided by ontologies available on the Semantic Web. Cattuto et al. [4] proposed a systematic evaluation of similarity measures for folksonomies. Voss [6] concentrates on the description of an approach for the translation of folksonomies to Linked Data and SKOS. Comparatively, most of the previous works concentrate on the analysis and alignment of simple (non-complex) tags. Another difference is the proposal of a representation model which goes beyond a taxonomic structure.

3   Representation Model

The representation model is aimed towards facilitating the fine-grained integration between different NLCDs, providing the creation of an integrated and more structured model from the category descriptors. The representation also has an associated interpretation model which aims at making explicit the algorithmic interpretation of the descriptor in the integrated graph. A NLCD can be segmented into 7 representation elements:

---

[1] http://graphia.dcc.ufrj.br/nlcd

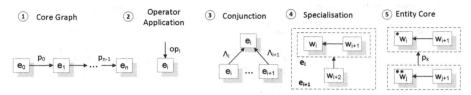

Fig. 1. Graph patterns showing the relations present in the graph representation

Entity: Entities inside a NLCD are terms which are sub-expressions of the original category which can describe predications or individuals. The entities map to a subset of the content words (open class words), which carry the main content or the meaning of a NLCD. Words describing entities can combine *nouns, adjectives* and *adverbs*. The entities for an example NLCD *'Snow Or Ice Weather Phenomena'* are *'Snow'*, *'Ice'*, *'Weather Phenomena'*. Entities are depicted as e_i in Figure 1(1).

Class & Entity core: Every entity contains a semantic nucleus, which corresponds to the phrasal head and which provides its core meaning. For the predicate *'Snow Or Ice Weather Phenomena'*, *'Phenomena'* is the class & entity core. Depicted as '*' in Figure 1(5).

Relations: Relation terms are binary predicates which connect two entities. In the context of predicate descriptors, relation terms map to closed class words and binary predicates, i.e. prepositions, verbs, comparative expressions (*same as, is equal, like, similar to, more than, less than*). Depicted as p_i in Figure 1(1).

Specialization relations: Specialization relations are defined by the relations between words w_i in the same entity, where w_{i+1} is specialised by w_i. Represented by an unlabelled arrow in Figure 1(4).

Operators: Represents an element which provides an additional qualification over entities as a unary predicate. Operators are specified by an enumerated set of terms which maps to adverbs, numbers, superlatives, etc. Depicted in Figure 1(2).

Conjunctions & Disjunctions: A disjunction between two elements ($w_i \lor w_{i+1}$) over an element e_j is defined by the distribution of specialization relations: e_j is specialised by w_i and e_j is specialized by w_{i+1}. A conjunction is treated as an entity which names the conjunction of two entities through a conjunction labelled link. The conjunction representation is depicted in Figure 1(3).

Temporal Nodes: Consists in the representation of temporal elements references into a normalized temporal range format.

Further examples are depicted in Figure 2. The representation can be directly translated into an RDF (Resource Description Framework) graph. Most of the overhead in the translation is due to the fact that words mapping to classes need to be instantiated and later reified. Terms which are classes and which need to be reified are reflected as instances.

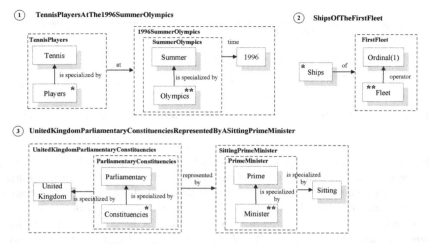

Fig. 2. Depiction of examples of NLCDs

4 Extraction

This section describes the process for extracting NLCDs into the proposed representation model. Figure 3 shows the components and the extraction workflow. The NLCD extraction consists of the following steps:

POS Tagging: Detection of the lexical categories of the NLCD words. The extractor uses the NLTK POS Tagger[2].
Segmentation: The segmentation of the NLCD starts by detecting the relations and splitting the descriptor into a set of entities and relations.
Entity Detection: This step consists on the detection of 3 types of entities: named entities, operators and temporal references: (i) The detection of named entities is based on the creation of a gazetteer from DBpedia 3.9 instances. Elements tagged as nouns and proper nouns are checked against the gazetteer; (ii) Operators are detected using the combination of an enumerated list of operators and regular expressions based on POS Tags; (iii) Temporal references are detected using regular expressions and are normalized.
Specialization ordering: This step consists in defining the specialization sequence for the terms inside each entity. Two heuristic indicators are used in the determination of the ordering of the terms inside the classes: POS Tags and a corpus-based specificity measure (inverse document frequency (IDF) over Wikipedia 2013 text collection). The POS Tags are used to order the words based on the lexical categories. The ordering is defined by the relations (NN - *is specialised by* → JJ, JJ - *is specialised by* → RB). For an entity containing words from the same lexical category, IDF is used to define the ordering: Lower IDF - *is specialised by* → Higher IDF.

---

[2] http://www.nltk.org

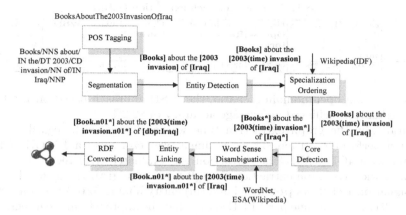

Fig. 3. Extraction components and workflow

Word Sense Disambiguation (WSD): The WSD component is used to align the extracted words with their WordNet senses, based on the context in which the word occurs (the NLCD). Let the sequence of words $w_0, w_1, ..., w_n$ be the natural language descriptor for a category c. Let $g_0, g_1, ..., g_k$ be the WordNet glosses associated with the senses for w_i for $0 \leq i \leq n$. Let $\kappa(w_i)$ be the context of w_i defined by $w_0, w_1, ..., w_n$ (excluding the target word). The sense for w_i is given by $sr_{ESA}(\kappa(w_i), g_j)$ where sr_{ESA} is the distributional semantic relatedness measure (Explicit Semantic Analysis) between the WordNet glosses and the category context.

Entity linking: The Entity linking component aligns terms in the extracted graph to DBpedia entities. Entity linking uses DBpedia as a named entity base and a ranking function based on TF/IDF over labels, entity cardinality and levehnstein distance.

RDF conversion: At this point the relations are represented as an internal set of extracted graphs following the proposed representation model. The model is then converted into an RDF graph.

5   Evaluation

The extraction approach was evaluated by randomly selecting a sample of 2,696 Wikipedia categories from the original set of 287,957 categories. These categories were extracted and manually evaluated according to eight extraction features (Table 1). The features map to the components of the extraction approach. Table 1 shows the accuracy for each feature.

The low error in entity segmentation, relation extraction and specialization sequence shows the generality of the extraction rules in relation to the tractable subset of natural language category descriptors. Additionally, the high accuracy in the determination of the sequence of specialization relations, detection of class and entity cores shows the correctness in the construction of the taxonomic

Table 1. Accuracy for each extraction feature

	Entity Segmentation	Relations	Unary Operators	Specialization Relations	Class Core	Entity Core	WSD	Entity Linking
Accuracy	79.38%	95.96%	99.74%	97.81%	99,37%	81.86%	82.2%	78.1%

structure. For an open domain scenario, the WSD approach based on Explicit Semantic Analysis achieved an average accuracy of 82,2%.

Additionally, the graph extraction time was evaluated with regard to the extraction performance time. The experiment was carried in a 1.70GHz CPU computer with 4GB RAM. The extraction was evaluated with regard to three main categories: (i) graph extraction time (**9.8 ms per graph**), (ii) word sense disambiguation **121.0 ms per word** and (iii) entity linking **530.0 ms per link**. The overall extraction time per NLCD shows that the approach can be integrated into medium-large scale categorization tasks. Each category generates an average of 10.2 RDF triples. The extraction tool is available at the website[3].

6   Conclusion and Future Work

This paper analyses the use of complex natural language category descriptors (NLCDs) and proposes a representation model and an extraction approach for NLCDs. The accuracy of the proposed approach was evaluated over Wikipedia category links, achieving an overall structural accuracy above 78%. Future work will focus on the evaluation of the approach under domain-specific NLCDs.

References

1. Suchanek, F., Kasneci, G., Weikum, G.: YAGO: A Large Ontology from Wikipedia and WordNet. In: Proc. of the 16th Intl. Conf. on World Wide Web, pp. 697–706 (2007)
2. Gabrilovich, E., Markovitch, S.: Computing semantic relatedness using Wikipedia-based explicit semantic analysis. In: Proc. of the Intl. Joint Conf. on Artificial Intelligence (2007)
3. Specia, L., Motta, E.: Integrating Folksonomies with the Semantic Web. In: Franconi, E., Kifer, M., May, W. (eds.) ESWC 2007. LNCS, vol. 4519, pp. 624–639. Springer, Heidelberg (2007)
4. Cattuto, C., Benz, D., Hotho, A., Stumme, G.: Semantic grounding of tag relatedness in social bookmarking systems. In: Sheth, A.P., Staab, S., Dean, M., Paolucci, M., Maynard, D., Finin, T., Thirunarayan, K. (eds.) ISWC 2008. LNCS, vol. 5318, pp. 615–631. Springer, Heidelberg (2008)
5. Limpens, F., Gandon, F., Buffa, M.: Linking Folksonomies and Ontologies for Supporting Knowledge Sharing: a State of the Art. Technical Report (2009)
6. Voß, J.: Linking Folksonomies to Knowledge Organization Systems. In: Dodero, J.M., Palomo-Duarte, M., Karampiperis, P. (eds.) MTSR 2012. CCIS, vol. 343, pp. 89–97. Springer, Heidelberg (2012)
7. Cimiano, P., Handschuh, S., Staab, S.: Towards the Self-Annotating Web. In: Proc. of the 13th Intl. Conf. on World Wide Web, pp. 462–471 (2004)

---

[3] http://graphia.dcc.ufrj.br/nlcd

Semantic and Syntactic Model of Natural Language Based on Tensor Factorization

Anatoly Anisimov, Oleksandr Marchenko, Volodymyr Taranukha,
and Taras Vozniuk

Faculty of Cybernetics, Taras Shevchenko National University of Kyiv, Ukraine

Abstract. A method of developing a structural model of natural language syntax and semantics is proposed. Factorization of lexical combinability arrays obtained from text corpora generates linguistic databases that used for natural language semantic and syntactic analyses.

1 Introduction

Recently, a non-negative tensor factorization (NTF) method has become widely used in the natural language processing. From among numerous works in the area of particular interest are two works [1, 2]. They describe models for the tensor representation of the frequency with which various types of syntactic word combinations can occur in sentences.The N-dimensional tensors contain estimates of the frequency of word combinations sets in sentences of texts corpora. After NTF such a model allows for successful automatic extraction of specific linguistic structures from a corpus, such as selectional preferences [1] and Verb Sub-Categorization Frames [2], which combine data on syntactic and semantic properties of relations between verbs and their noun arguments in sentences. The number of dimensions in the tensor restricts the maximum length of sentences and phrases described by this model. Van de Cruys describes a three-dimensional tensor for modeling the syntactic combination: Subject -Verb - Object [1]. Subsequently, Van de Cruys and colleagues describe tensors of 9 and 12 dimensions to simulate about twenty different types of syntactic relations [2]. The mere increase in the tensor dimension number, however, does not seem to be a good way of improving the model and handling more types of complex syntactic relations. It is quite reasonable to look for other universal representation models for syntactical structures. The control spaces [3] have been chosen from among numerous time-tested classical formal models of language syntax representation owing to the fact that in this model an arbitrary complex structure is described using recursion through superposition of two basic syntactic relationships - syntagmatic and predicative. The proposed here lexical and syntactic tensor model consists of a 3-dimensional tensor for predicative relations (like Subject - Verb - Object) and a matrix for syntagmatic relations (like Noun -Ajective, Verb -Adverb, etc.). Sentences have two types of links: a strictly linear relation (predicative structure) and a closed cyclic dependency (syntagmatic definition). The use of control spaces appears to be an effective means to reduce arbitrary n-ary syntactic relation to the superposition of binary and ternary relations.

E. Métais, M. Roche, and M. Teisseire (Eds.): NLDB 2014, LNCS 8455, pp. 51–54, 2014.

2   Lexical-Syntactical Model of a Natural Language

In order to construct a semantic-syntactic model of a natural language, method for automatic filling the three dimensional tensor F(for linear predicative relations) and the matrix D (for cyclic syntagmatic dependencies) was designed. The method calls for the following steps:

- Sentences from a large text corpus are taken and parsed by the Stanford Parser module, which generates the syntactic structures of sentences in the form of dependency trees and parse trees for CF phrase structure grammar;
- The program examines the dependency tree and the CFG parse tree of the current sentence, while constructing the control space of syntactic structure, analyzing relations between corresponding words to identify predicate combinations of length 3 (e.g., Subject-Verb-Object, etc.) and syntagmatic combinations of length 2 (Noun-Adjective, Verb-Adverb, etc.);
- In the control space of this sentence for every triad of points (i, j, k) connected with the linear predicative sequence of links, in the tensor F for the cell $F[I, J, K]$ receives the value: $F[I, J, K] = F[I, J, K] + 1$. The coordinates I, J, K of the tensor cell correspond to pairs (w_i, A_i), (w_j, A_j) and (w_k, A_k), where w means words that are lexical values of the corresponding points (i, j, k), and A is a coded description of the characteristics of these words (part of speech, gender, number of lexical units, etc.).
- Similarly, in the control space of the syntactic structure of the current sentence for each pair of points (i, j) interconnected with the cyclic syntagmatic link, in the matrix D for the cell $D[I, J]$ is set to: $D[I, J] = D[I, J] + 1$.

The extremely large dimension and sparsity of the matrix D and the tensor F demand non-negative matrix factorization (NMF) and NTF in order to store the data in a more economical way. Factorization of the matrix D is performed using Lee and Seung NMF algorithm that decomposes the matrix $D(N \times M)$ as a product of two matrices $W(N \times k) \times H(k \times M)$, where $k << N, M$. Factorization of the tensor F is performed using the NTF parallel algorithm PARAFAC. Three corresponding matrices X, Y and Z have been generated.

After the matrix D and the tensor F factorization, the system forms a strong knowledge base which contains information about the framework of syntactic structures of natural language sentences. In order to determine whether two words a and b form a ringed syntagmatic relation, one has to take the vector-row W_a from the matrix W corresponding to the word a, and the vector-column H_b from the matrix H which corresponds to the word b, and calculate the scalar product of the vectors (W_a, H_b^T). If the product is greater than a certain threshold T, then this relation is defined. In order to determine whether the three words a, b and c form a predicative relation $(a \rightarrow b \rightarrow c)$, it is necessary to take the vector X_a corresponding to the word a, the vector Y_b corresponding to the word b, and the vector Z_c corresponding to the word c and to calculate the value: $S_{abc} = \sum_{i=1}^{k} X_a[i] * Y_b[i] * Z_c[i]$. If S_{abc} value is greater than the threshold, then this relation is defined. If not, it is considered undefined.

3   Application and Experiments

As the initial training text corpus, sets of articles from the English Wikipedia (EW), the Simple English Wikipedia (SEW) and The Wall Street Journal (WSJ) corpus are used. At first, the sentences are analyzed with the Stanford parser yielding a CFG parse tree and a dependency tree. Also, the algorithm has been developed to construct control spaces by converting the dependency tree and the CFG parse tree into the control space of a sentence. 800,000 articles from the EW and the SEW have been processed, along with the WSJ corpus. The processing yielded the large matrix D for ringed links (numbering approximately 2.3 million words $\times$ 2.3 million words, with about 57 million nonzero elements) and the large three-dimensional tensor F for linear predicative connections (consisting of approximately 2.3 million words $\times$ 52 thousand words $\times$ 2.3 million words, with about 78 million non-zero elements). These arrays were factorized. A parser for the English language based on the obtained arrays of lexical-syntactic combinability was implemented. This parser, based on the Cocke-Younger-Kasami algorithm, directly constructs the control space of a sentence. The model describes only the relations among those words which actually occur in the corpus sentences and have been processed accordingly. This can be easily dealt with using synonym dictionaries. In the system we developed the WordNet is used to this end. We assume that if a relation exists between A and B, it also exists between an arbitrary pair of A_i and B_i, where A_i is any word from the synset that contains A, while B_i is any word from the synset that contains B. But many words refer to several different synsets with different meanings. There are several approaches to solve the problem of ambiguous words. The two matrices W and H resulting from the NMF of matrix D can be considered as powerful tools for determining the degree of semantic similarity between words according to the latent semantic analysis method. To determine the presence of ringed syntagmatic connections between the word a and the word b and to solve the problem of ambiguous words the following steps are carried out:

A: Take the vector W_a corresponding to the word a from the term matrix W, the vector column H_b which corresponds to the word b from matrix H, and calculate the scalar product of the vectors (W_a, H_b^T). If the value $(W_a, H_b^T) > T$, then this link is **defined**. T is the threshold. The optimal value of T is found experimentally. If it fails:

B: Take synsets for the words a and b from the WordNet. The set of synsets $\{A_i\}$ refers to the word a, and the set of synsets $\{B_i\}$ refers to the word b. Check the pairs of the words formed from the elements of $\{A_i\}$ and $\{B_i\}$. If there is the word a'_k from $A_k \in \{A_i\}$ and the word b'_j from $B_j \in \{B_i\}$ such that scalar product of vectors $(W_{a'_k}, H_{b'_j}^T) > T$, then this link between a and b is **defined**. If not:

C: The set $\{A_i\}$ is expanded with synsets linked with nodes from $\{A_i\}$ with hyponym and hypernym relations in the WordNet. The set $\{B_i\}$ is expanded in the same way. Check the pairs of words formed from elements of $\{A_i\}_{exp}$ and $\{B_i\}_{exp}$ (excluding already checked pairs on step B). If there is the word a'_k from the synset $A_k \in \{A_i\}_{exp}$ and the word b'_j from the synset $B_i \in \{B_i\}_{exp}$ such that the scalar product of vectors $(W_{a'_k}, H^T_{b_j}) > T$, then the link between a and b is **defined**. If it fails: expand $\{A_i\}_{exp}$ and $\{B_i\}_{exp}$ recursively 2 or 3 times and repeat step (C).

If it is always $(W_{a'_k}, H^T_{b_j}) < T$, then the link **does not exist**.

For the linear predicative links this algorithm works in the same way.

The experiments were performed on the WSJ corpus using cross-validation. The test on the WSJ corpus was performed automatically. It was carried out taking into account the algorithmic case in which a particular syntactic relation was found. The comparative analysis was performed only for the sentences that had been successfully processed by the developed parser with the complete building of the syntactic structures (92.7 % sentences were successfully processed for the WSJ corpus). The results are summarized in Table 1.

Table 1. Estimation of the accuracy of building ringed-syntagmatic links and linear predicative links on sentences from the WSJ corpus

Algorithmic case	Ringed-syntagmatic link	Linear predicative-links
Case A	93,71%	94,37%
Case B	91,05%	91,33%
Case C	85,07%	89,79%

4   Conclusions

The recursiveness of syntactic control spaces allows us to describe sentence structures of arbitrary complexity, length and depth. This enables the development of a semantic-syntactic model based on a single three-dimensional tensor and a single matrix instead of increasing the number of dimensions of connectivity arrays for lexical items. The system for analysis and constructing control spaces has been developed on the basis of factorized arrays. It shows high quality and accuracy, thus proving the correctness and effectiveness of the constructed model.

References

1. Van de Cruys, T.: A Non-Negative Tensor Factorization Model for Selectional Preference Induction. Journal of Natural Language Engineering 16(4), 417–437 (2010)
2. Van de Cruys, T., Rimell, L., Poibeau, T., Korhonen, A.: Multi-Way Tensor Factorization for Unsupervised Lexical Acquisition. In: Proceedings of COLING 2012, pp. 2703–2720 (2012)
3. Anisimov, A.V.: Control Space of Syntactic Structures of Natural Language. Cybernetics and System Analysis 3, 11–17 (1990)

How to Populate Ontologies

Computational Linguistics Applied to the Cultural Heritage Domain

Maria Pia di Buono, Mario Monteleone, and Annibale Elia

University of Salerno, Fisciano (SA), Italy
{mdibuono,mmonteleone,elia}@unisa.it

Abstract. The Cultural Heritage (CH) domain brings critical challenges as for the application of Natural Language Processing (NLP) and ontology population (OP) techniques. Actually, CH embraces a wide range of content, variable by type and properties and semantically interlinked whit other domains. This paper presents an on-going research on language treatment based on Lexicon-Grammar (LG) approach for improving knowledge management in the CH domain. We intend to show how our language formalization technique can be applied for both processing and populating a domain ontology.

Keywords: NLP, Ontology Population, Cultural Heritage.

1   Introduction

Semantic expansion is a specific characteristic of the CH domain. This concept includes different aspects, which regard tangible and intangible heritage, and generates several kinds of descriptive metadata. In this perspective, NLP techniques for text extraction and mining are an attempt for bridging the information gap and improving access to cultural resources. Therefore, this paper presents on-going research on language treatment based on LG approach for improving knowledge management in the CH domain. We intend to show how our language formalization technique can be applied for both processing and populating a domain ontology. Indeed, according to Bachimont [1], "defining an ontology for knowledge representation tasks means defining, for a given domain and a given problem, the functional and relational signatures of a formal language and its associated semantics".

2   Methodology

2.1   LG Framework

LG is our NLP theoretical and practical framework. It formalizes natural language in order to achieve text automatic parsing, describing all mechanisms of word combinations and giving an exhaustive description of natural language lexical and syntactic

E. Métais, M. Roche, and M. Teisseire (Eds.): NLDB 2014, LNCS 8455, pp. 55–58, 2014.

structures. Set up by the French linguist Maurice Gross [2] during the '60s, LG was subsequently applied to Italian by Annibale Elia, Maurizio Martinelli and Emilio D'Agostino [3]. Theoretically it is based on Zelig Harris' [4] Operator-Argument Grammar, for which each human language is a self-organizing system where word syntactic and semantic properties may be calculated on the basis of the relationships they have with all other co-occurring words inside nuclear or simple sentence contexts. LG mainly studies simple sentences analyzing distributional and transformational rules based on predicate syntactic-semantic properties (i.e. co-occurrence and selection restriction rules). In LG, also electronic dictionaries are used to describe words morph-grammatical features. These dictionaries are mainly based on the concepts of «meaning unit», «lexical unit» and «word group», this last one also including Multi Word Units (MWUs) and terminological compound words, which even being very often semantically compositional, can be lemmatized due to their particular non-ambiguous information content [5]. Also, we use co-occurrence and selection restriction to build NLP tools called local grammars, and which can parse specific linguistic patterns in natural language-based applications.

2.2   Formal Structure of LG Electronic Dictionaries of Compound Words

Our CH electronic dictionary are based on the Thesauri and Guidelines of the Italian Central Institute for the Catalogue and Documentation (ICCD)[1]. Together with morph-grammatical and formal descriptions, to each entry an ontological identification is given, consisting in tags which send back to the knowledge domain within which they are commonly used with terminological non-ambiguous meanings. The following list is an example of the Italian Electronic Dictionary of the Archaeological Domain:

> freccia di balestra,N+NPN+FLX=C45+DOM=RA1SUOARAL
> freccia foliata,N+NA+FLX=C556+DOM=RA1SUOIL
> fregio con coronamento,N+NPN+FLX=C12+DOM=RA1EDEAES
> fregio dorico,N+NA+FLX=C523+DOM=RA1EDEAES

In the compound word *fregio dorico* (Doric frieze) the tag «NA» indicates that the given compound is formed by a Noun, followed by an Adjective. The tag «+FLX=C523» indicates the inflectional class refers to a local grammar in order to retrieve all feasible compound forms. The compound word is also marked with the tag «+DOM=RA1EDEAES», which stands for «Archaeological Artefacts – Building – Architectural Elements – Structural Elements» and indicates the taxonomy class. We also preview domain label subset in order to indicate specific subsectors. For Archaeological Artefacts domain, in which a generic tag «RA1» is used, more explicit tags are used for Object Type, Subject, Primary Material, Method of Manufacture, Object Description. Currently the Italian electronic dictionary for the Archaeological Domain is composed by about 11000 tagged compound words. According to our

---

[1] http://www.iccd.beniculturali.it/index.php?it/240/vocabolari

approach, electronic dictionaries entries (simple words and MWUs) are the subject and the object of the RDF triple.

2.3   Finite State Automata/Finite State Transducers (FSA/FSTs)

Finite-state transducers (FSTs) and finite-state automata (FSA) are here used to build local grammars. An FST is a graph representing a set of text sequences and associating each recognized sequence to specific analysis results. Text sequences are described in the FST output. An FSA is a special type of FST which has no output [6]; it is typically used to locate morph-syntactic patterns inside corpora. It extracts matching sequences in order to build indices, concordances, etc.

Fig. 1. Example of a simple FSA for the recognition of Archaeological artifacts descriptions

Fig. 1 shows an FSA composed of a single path with four nodes (from the initial symbol on the left to the end symbol on the right). Words in angle brackets stand for lemma forms (generally all inflected, derived forms, or spelling variants of a given lexical entry), the shaded boxes represent a sub-graph which can freely be embedded in more general graphs. At a more theoretical level, it introduces the power of recursion inside grammars. Sub-graphs may also be used to represent a semantic class and can be encoded in a dictionary with specific semantic features. When the word form is set between angle brackets, the software locates all the word forms that are in the same equivalence set as the given word form.

3   Semantic Annotation and Ontology Population

We rely upon the ontology given by the International Council of Museums - Conseil Interational des Musees (ICOM – CIDOC) Conceptual Reference Model (CRM) [7]. It is composed of 90 classes (which includes subclasses and super-classes) and 148 unique properties (and sub-properties). This object-oriented semantic model is compatible with the Resource Description Framework (RDF). We use FSA variables for identifying ontological classes and properties for subject, object and predicate within RDF graphs. In Fig. 2 we develop an FSA with variables which apply to the Parts of Speech (POS) CIDOC CRM classes and properties: (i) E19 indicates "Physical Object" class, (ii) P56 stands for "Bears Feature" property, (iii) E26 indicates "Physical Feature" class. The role pairs Physical Object/name and Physical Feature/type are trigged by the RDF predicate *presenta*.

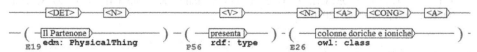

Fig. 2. Sample of the use of the FSA variables for identifying classes and property

In Fig. 2 we used variables to apply a tag which indicates classes for the compound words; when the FSA recognizes the text string, it applies a text annotation. Besides, in Fig. 2 we also indicate specific Parts of Speech (POS) for the first noun phrase *Il Partenone* (DETerminer + Noun), the verb *presenta* (V) and the second noun phrase *colonne doriche e ioniche* (Noun+Adjective+Conjuntion+Adjective). By applying the automaton in Fig. 2 (built using the high variability of lexical class and not of the original form) we can recognize all instances included in E19 and E26 classes, the property of which is P56. This matching of linguistic data to RDF triples and their translation into SPARQL/SERQL path expressions allows the use of specific meaning units to process natural language queries.

4   Conclusion

Our methodology relies on a linguistic processing phase requiring robust resources and background knowledge, but it allows to perform both object/term and synonym identification and also to recognize relations. As such, it can also ensures portability to other domains, preserving ontology consistency and entity disambiguation. This means that NLP routines based on LG allow to support the automatic semantic annotation/indexation of CH textual documents. Our future goal is the development, inside NOOJ[2], of a module for semi-automatic OP. Starting from the entries retrieved and from their specific tags, stored in electronic dictionaries and in FSA/FSTs, such tool will write and fill all fields directly using RDF schema and OWL, automatically generating strings while correctly coupling ontologies and compound words.

References

1. Bachimont, B.: Engagement sémantique et engagement ontologique: conception et réalisation d'ontologies en ingénierie des connaissances. In: Charlet, J., Zacklad, M., Kassel, G., Bourigault, D. (eds.) Ingénierie des Connaissances, Évolutions Récentes et nouveaux défis. Eyrolles, Paris (2000)
2. Gross, M.: Méthodes en syntaxe. Hermann, Paris (1975)
3. Elia, A., Martinelli, M., D'Agostino, E.: Lessico e strutture sintattiche. Introduzione alla sintassi del verbo italiano. Liguori Editore, Napoli (1981)
4. Harris, Z.S.: Notes du cours de syntaxe, traduction française par Maurice Gross. Le Seuil, Paris (1976)
5. Vietri, S., Monteleone, M.: The English NooJ dictionary. In: Proceedings of NooJ 2013 International Conference, June 3-5, Saarbrücken (2013) (in press)
6. Silberztein, M.: Dictionnaires électroniques et analyse automatique de textes. Masson, Paris (1993)
7. Crofts, N., Doerr, M., Gill, T., Stead, S., Stiff, M.: Definition of the CIDOC Conceptual Reference Model. ICOM/CIDOC Documentation Standards Group.CIDOC CRM Special Interest Group. 5.02 ed. (2010)

---

[2] For more information on NooJ, see www.nooj4nlp.org

Fine-Grained POS Tagging of Spoken Tunisian Dialect Corpora

Rahma Boujelbane, Mariem Mallek, Mariem Ellouze,
and Lamia Hadrich Belguith

ANLP Research Group, MIRACL Lab., University of Sfax, Tunisia
rahma.boujelbane@gmail.com, mariem.mallek@hotmail.fr,
{Mariem.Ellouze@planet.tn}@planet.com, l.belguith@fsegs.rnu.tn

Abstract. Arabic Dialects (AD) have recently begun to receive more attention from the speech science and technology communities. The use of dialects in language technologies will contribute to improve the development process and the usability of applications such speech recognition, speech comprehension, or speech synthesis. However, AD faces the problem of lack of resources compared to the Modern Standard Arabic (MSA). This paper deals with the problem of tagging an AD: The Tunisian Dialect (TD). We present, in this work, a method for building a fine grained POS (Part Of Speech tagger) for the TD. This method consists on adapting a MSA POS tagger by generating a training TD corpus from a MSA corpus using a bilingual lexicon MSA-TD. The evaluation of the TD tagger on a corpus of text transcriptions achieved an accuracy of 78.5%.

Keywords: Spoken corpus, POS tagging, Tunisian Dialect.

1   Introduction

The Arabic language has multiple variants, including Modern Standard Arabic (MSA), the formal written standard language of education, and the informal spoken AD. In fact, AD are the true native language forms. They are generally restricted in use to the informal daily communication. They are not taught in schools or even standardized. They are spoken but not written. To date, AD is emerging as the language of the news and of many varieties of television programs, and also of informal communication online, in emails, blogs, discussion forums, chats, SMS, etc. Although there are commercially Natural Language Processing (NLP) systems for processing MSA with low error rates, these systems fail when an AD is introduced. That's why there is an inherent need for large scale annotated resources for dialects. In this study, we focus on creating annotated resources for an AD: the TD. Our aim in this paper consists in adapting an MSA POS tagger to TD, espacially to the dialect spoken in TV and radio broadcasts in which speakers often code switch between MSA and dialect. However, we lack resources, TD tagged corpora, for training the tagger. In fact, several researchers have explored the idea of exploiting existing MSA rich resources to build tools for DA (Dialect arbic) NLP. So, we developed a tool that generate, from an

E. Métais, M. Roche, and M. Teisseire (Eds.): NLDB 2014, LNCS 8455, pp. 59–62, 2014.

MSA corpora, a TD POS tagged corpora [1]. Thereafter, we explore a bilingual lexicon MSA-TD for building rules to enrich POS tags by other morphological features. This paper is organized as follows: In section 2, we present the TD and we outline works done on it. In section 3, we present our method for building a fine-grained POS tagger for the spoken Tunisian corpora. Finally, in Section 4 we discuss the results.

2   Tunisian Dialect

The TD is an AD attached to the Arab Maghreb and is spoken by twelve million people living mainly in Tunisia. It is generally known to its speakers as the 'Darija' or 'Tounsi', which simply means "Tunisian", to distinguish it from Modern Standard Arabic. The TD is considered an under'-'resourced language. It has neither a standard orthographic or written text nor dictionaries. Actually, there is no strict separation between Modern Standard Arabic (MSA) and its dialects, but a continuum dominated by mixed forms (MSA-Dialect). Many researchers in the ANLP research group[1] focused on the automatic processing of the TD. In fact, Graja et al. [2], treated the TD for understanding speech. To do so, the researchers relied on manual transcripts of conversations between agents at the train station and travelers. Boujelbane et al. [1] built lexical resources such as a bilingual lexicon MSA-TD for verbs, nouns and tool words to convert MSA corpora to TD corpora. Hamdi et al. [3] adapted Magead (analyzer and generator of MSA verbs) [4] to analyze and generate TD verbs by using the bilingual lexicon of verbs. Since the TD corpora and dictionaries should respect a standard orthography, Zribi et al [5] built a Tunisian CODA (Conventional Orthography for Dialectal Arabic) which is very helpful to build standard resources (corpora, transcriptions, dictionaries...).

3   Adapting an MSA POS Tagger to the TD

Morpho-syntactic tagging is essential, as a preliminary step to any high level processing the text. Different reliable taggers exist for MSA, but they have been designed for handling written texts and are, therefore, not suited for less normalized language like TD language. So, to tag a TD spoken corpus with POS tags, we proposed to re-use an MSA POS tagger:The Stanford POS Tagger [6]. the state of the art has shown that tagging methods based on discriminative probabilistic models reach very high levels of accuracy on Arabic (95.49% with SVM and 96.5% with Maximum-Entropy MaxEn).

3.1   Building a Training Corpus

As we have mentioned, the TD is considered as an under-resourced language. There is no tagged Tunisian text. Thus, we explored the idea that resources available to tag MSA can be advantageously used to create dialectal resources.

---

[1] https://sites.google.com/site/anlprg/

So, as the Stanford MSA tagger was trained on the ATB corpus [7], we generate a TD version of the ATB using our TDT translator tool [1]. The TDT annotates each MSA text morpho-syntactically by using MADA analyzer (Morphological Analyzer and Disambiguator of Arabic) [8]. Then, based on each part of speech of the MSA-word, the TDT proposes for each MSA structure the corresponding TD translation with the appropriate POS and this by exploring the bilingual dictionaries MSA-TD.

3.2 POS Tag's Enrichement

Many computational applications use more fine-grained POS tags for processing languages. That's why we propose to enrich the TD POS tags by other morphological features. To do this, we proceeded as follow: First, we define the set of TD affixional morphemes and removed them from the word. For example, we remove from the noun ¡hadafuhum¿/*hadafuhum*/their aims the suffix ¡hum¿/ *hum*/their. After removing affixes, we transforme each stem to its pattern form; for example, the word ¡hadaf¿/ɪhadaf/event becomes CVCVC (Consonant Vowel Consonant Vowel Consonant). Then, we deduct from the lexicon rules that describe the morphological analyses of each variation of a pattern. If the scheme of the stem is found among the set of patterns, we can define the morphological features for the word; for example: the word ¡Hadaf¿ will have these features: PAT: CaCaC, Gender: MASC, Number: SG, Root: Hdf ¡hadaf¿ Finally, we re-add the appropriate affixes to the stem and we enrich the tags by their morphosyntactic properties.

4 Evaluation

To evaluate the performance of the tagger to tag a spoken corpus with POS,we transcribed 2h15min of news (4041 words) broadcasted by a Tunisian channel and we annotated them manually using the same tags as the training corpus. These transcriptions mix MSA and TD in their speech. So, First, we annotate the transcriptions through Stanford MSA (baseline). It attained an accuracy

Table 1. The TD-Stanford POS tagger evaluation

	TD-Stanford			MSA-stanford
	recall	precision	accuracy	accuracy
Verbs	77%	55%	78,5 %	63,5%
Nouns	87%	89%		
Adjectives	95%	72%		
Adverbs	46,5%	100%		
Pronouns	71,5%	100%		
Conjunction	14%	83%		
Particles	57%	96%		
Preposition	67%	100%		
Propoer nouns	65%	84%		

of 63.5%. Then, we annotate the same transcriptions using the TD version of Stanford. It achieved an accuracy of 78.5%. The increase of accuracy shows that the conversion MSA-TD using the bilingual lexicon is useful to adapt MSA tools to TD. Table 1 show the performance of the TD Stanford tagger.

5   Conclusion

The development of tagged corpora is, definitely, a human resource and time-consuming task. In this paper, we presented a process for an atomatic tagging of the spoken TD dialect. The process of tagging a spoken Tunisian corpus through the reuse of already available resources which is presented may constitute an example of how to minimize the cost of such task without compromising the results. To our knowledge, the tagger which has been developed is the first one for the TD. In a future work, we aim to enhance the tagger's accuracy by increasing the size of the lexicon. We plan also to enhance it so that it will be able to identify whether the word is in TD or in MSA.

References

1. Boujelbane, R., Khemekhem, M.E., Belguith, L.H.: Mapping Rules for Building a Tunisian Dialect Lexicon and Generating Corpora. In: Proceeding of International Joint Conference on Natural Language Processing (IJCNLP), Nagoya, Japan (2013)
2. Graja, M., Jaoua, M., Belguith, L.H.: Towards Understanding Spoken Tunisian Dialect. In: Lu, B.-L., Zhang, L., Kwok, J. (eds.) ICONIP 2011, Part III. LNCS, vol. 7064, pp. 131–138. Springer, Heidelberg (2011)
3. Hamdi, A., Boujelbane, R., Habash, N., Nasr, A.: Un systme de traduction de verbes entre arabe standard et arabe dialectal par analyse morphologique profonde. Traitement Automatique des Langues Naturelles (2013)
4. Habash, N., Rambow, O., Kiraz, G.: Morphological analysis and generation for Arabic dialects. In: Proceedings of the ACL Workshop on Computational Approaches to Semitic Languages (2005)
5. Zribi, I., Boujelbane, R., Masmoudi, A., Khemakhem, M.E., Belguith, L., Habash, N.: A Conventional Orthography for Tunisian Arabic. In: The Language Resources and Evaluation Conference (LREC), 9th edn., Iceland (2014)
6. Toutanova, K., Manning, C.D.: Enriching the knowledge sources used in a maximum entropy part-of-speech tagger. In: Proceedings of the 2000 Joint SIGDAT Conference on Empirical Methods in Natural Language Processing and Very Large Corpora: Held in Conjunction with the 38th Annual Meeting of the Association for Computational Linguistics (2000)
7. Maamouri, M., Bies, A., Buckwalter, T., Mekki, W.: The Penn Arabic Treebank: Building a Large-Scale Annotated Arabic Corpus. In: NEMLAR Conference on Arabic Language Resources and Tools, Cairo, Egypt (2004)
8. Habash, N., Rambow, O., Roth, R.: MADA+ TOKAN: A toolkit for Arabic tokenization, diacritization, morphological disambiguation, POS tagging, stemming and lemmatization. In: Proceedings of the 2nd International Conference on Arabic Language Resources and Tools (MEDAR), Cairo, Egypt (2009)

Applying Dependency Relations to Definition Extraction

Luis Espinosa-Anke and Horacio Saggion

Tractament Automàtic del Llenguatge Natural,
Department of Information and Communication Technologies,
Universitat Pompeu Fabra
{luis.espinosa,horacio.saggion}@upf.edu
http://taln.upf.edu

Abstract. Definition Extraction (DE) is the task to automatically identify definitional knowledge in naturally-occurring text. This task has applications in ontology generation, glossary creation or question answering. Although the traditional approach to DE has been based on hand-crafted pattern-matching rules, recent methods incorporate learning algorithms in order to classify sentences as definitional or non-definitional. This paper presents a supervised approach to Definition Extraction in which only syntactic features derived from dependency relations are used. We model the problem as a classification task where each sentence has to be classified as being or not definitional. We compare our results with two well-known approaches: First, a supervised method based on Word-Class Lattices and second, an unsupervised approach based on mining recurrent patterns. Our competitive results suggest that syntactic information alone can contribute substantially to the development and improvement of DE systems.

1 Introduction

Encyclopedias and terminological databases are knowledge bases of great importance for human and automatic processing, but their manual development is costly and slow [3]. Definition Extraction (DE), which is the task to automatically identify definitions in naturally-occurring text, can play an important role in this context. DE has received an increasing interest in Natural Language Processing, Computational Linguistics and Computational Lexicography. It has been proven effective for developing glossaries [16] [21], lexical databases [17] [34], developing question answering tools [25] [5], supporting terminology applications [13] [29] [28], automatically acquiring ontologies [39] or for developing e-learning applications [35] [8].

DE has been widely studied as a binary classification problem, where each sentence should be assigned the correct class, i.e. whether it is a definition. The majority of systems developed in the last few years addressing DE are based on hand-crafted rules aiming at identifying definitions in texts through pattern-matching, although a renewed interest in incorporating learning algorithms has emerged [7]. Following this line of work, we introduce a method for DE that

E. Métais, M. Roche, and M. Teisseire (Eds.): NLDB 2014, LNCS 8455, pp. 63–74, 2014.
© Springer International Publishing Switzerland 2014

exploits linguistically motivated features extracted from dependency relations. Benefiting from syntactic information extracted from dependency structures has proven efficient in other fields, like Information Extraction [32] [30] [1] or paraphrase identification [33]. With regard to DE, we are unaware of any supervised system which relies only on features derived from dependency relations, although these have been previously used, either for the extraction of definition candidates[9] [36] or as part of more comprehensive feature sets [6].

We evaluate the performance of our system and compare it with a supervised method based on word-class lattices [18] and with an unsupervised method that exploits recurrent patterns at various linguistic levels [38]. Moreover, our approach relies only on Machine Learning and skips the pattern-crafting stage, which seems to address issues like language-dependence and domain-specificity [7]. The main contributions of this work can be summarized as:

- A set of features based exclusively on dependency relations for identifying definitional statements in text.
- A set of experiments that demonstrate competitive performance when compared with state-of-the-art systems.

The remainder of this paper is divided as follows. Section 2 provides the reader with a general review of the most prominent work carried out in the field. Section 3 describes in detail the datasets used, the method followed for designing the features, and provides an illustrative example. Section 4 describes the experimental setup. Next, Section 5 covers the evaluation metrics and provides results with respect to comparable work in the field. Finally, Section 6 presents the main conclusions that can be drawn from the system described, as well as suggesting potential lines for future work.

2   Related Work

Traditionally, DE has been approached as a pattern-matching problem, where a set of hand-crafted rules are designed and applied to a dataset. One of the cue phrases that has been looked at has been the presence of specific trigger verbs [23] [25] [27] [31]. Moreover, other features like punctuation or layout features have also been considered in the pattern-matching process [16] [12] [26] [22] [15].

However, the increasing number of systems which are based totally or partially on Machine Learning reflect the current trend in DE. For instance, [9] exploit features like text, document and syntactic properties as well as named entity tags for training a classifier. [5] contributed with probabilistic lexico-syntactic patterns to modelling definitions. A soft matching model is described based on an n-gram language model. [4] use genetic programming to learn rules for typical linguistic forms in definitional sentences, and then genetic algorithms to weight these rules. Another generalizing approach relies on Word Class Lattices, which are generated from sentence clusters and applied to identify different types of definitions [18]. On the other hand, unsupervised pattern-mining approaches take into consideration sets of definitional sentences, which are extracted at three different levels (Ngram patterns, subsequence patterns, and dependency subtree patterns) [38]. Moreover, [24] compare two methods: First, a bootstrapping approach where lexico-syntactic patterns and candidates for definition terms are iteratively constructed, and second, a deep semantic analysis which exploits the semantic predicate-argument structure.

Finally, let us refer to recent work that addresses the DE problem not as a binary classification task, but as a sequential labeling task where each token is classified as being part of a definition or not [11] [8]. In both cases, the classification is performed using a probabilistic model based on Conditional Random Fields. However, these works differ in that the former aims at classifying different parts of a definition whereas the latter, although makes use of these features, does not explicitly tag these components.

3   Methodology

This section discusses the methodology followed for the implementation of our experiments. Firstly, the dataset used is described and, secondly, the linguistic motivation behind our data modelling is provided, followed by a description of the feature set.

3.1   Dataset

Our dataset consists of a collection of 1,908 definitional sentences and 2,711 non-definitional sentences extracted from Wikipedia [19]. Following the terminology generalization approaches from [18] and [38], all the definiendum[1] terms are reduced to the token *TARGET* in order to include this wildcard as a feature, thus assuming a prior step in terminology identification. In this context, a term is considered to be a word or phrase that has an entry in the English Wikipedia.

3.2   Modelling the Data

In the linguistic theory of Dependency Grammar, the notion of dependency is fundamental [20]. A syntactic structure is described by the distribution of lexical elements linked by asymmetrical relations called dependencies. One of the main characteristics is that, unlike constituent structures, a dependency tree has no phrasal nodes. Moreover, the dependency representations provide a direct encoding of predicate-argument structures (see the dotted dependencies in Figure 1). Finally, the relations between units in a dependency tree are bilexical, which makes them beneficial for disambiguation [20].

Within the dependency structure of a sentence, certain relations seem to be more informative in the context of DE. For example, sentences S1 and S2 below share the same surface structure (*TARGET is * which was *). However, only S1 is a definition [38].

---

[1] The *definiendum* refers to the term to be defined, while the *definiens* is a cluster of words that defines the term.

S1: *TARGET is the independent school which was opened.*
S2: *TARGET is secure against CCA2 which was proved.*

The main difference pointed out is that the dependency relation between the verb and the phrase "the independent school" is of **objective noun phrase**, while in S2, the relation with the adjective "secure" is of **adjectival phrase**. This is the kind of information that (in S1) an *Ngram* approach would be unable to tackle due to the non-adjacent distance between the components.

In the specific case of this paper, we consider two kinds of subtrees, which are: Parent-Child-Grandchild (PCG) (left) and Parent-Child-Child (PCC).

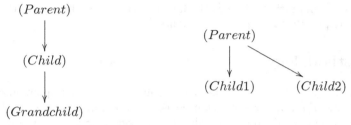

The PCG subtree allows extracting sequences like $Verb \rightarrow NN \rightarrow IN$, which in definitional sentences is an indicator of a hypernym being introduced and described (e.g. *The TARGET **is** a **family of** personal computers originally developed by Amiga Corporation*). This pattern accounts for 7% of all the 3-node subtrees in the definitional sentences present in the Wikipedia corpus, which can be considered to be high given the great sparseness of this set.

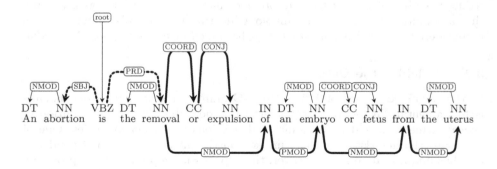

Fig. 1. Example sentence with dependency parsing

Syntactic dependency relations can reveal the domain or discipline governing a definition, and can be expressed by a locative at the beginning of a sentence. Consider the following sample sentence: ***In law**, an TARGET **is** a brief statement that contains the most important points of a long legal document or of several related legal papers.* The highlighted words $is(Verb, Root) \rightarrow In(Prep, Loc) \rightarrow law(NN, PMOD)$ form a $Parent \rightarrow Child \rightarrow Grandchild$ subtree, where the locative preposition *in* connects the topic with the verb. In the definitional section of the Wikipedia corpus, almost 20% of all the definitions have this structure at the beginning of the sentence.

Moreover, TARGET-is-a-*hypernym* patterns are a potential candidate for our subtree-based feature extractor. By attempting an ngram-based approach, we would be approaching the task as a surface pattern matching, perhaps including the Part-of-Speech of the candidate hypernym (to disregard, for example, noun phrases whose first word is an adjective). However, by looking at the syntactic function of the noun phrases involved (SBJ for TARGET, and Predicate for the hypernym), we can filter out some of the noisy subtrees retrieved.

With regard to the PCC subtree type, this can be useful for identifying *SVO* relations [30], as well as extracting multiword terminology. This can be further illustrated with the following sentence: *An TARGET is a **segmental writing system** which is based on consonants but in which vowel notation is obligatory.* The highlighted pattern follows a $(segmental, NN, NMOD) \leftarrow (system, NN, PRED) \rightarrow (writing, NN, NMOD)$ structure, and since we know that the word *system* is the syntactic predicate, we are highly confident that it can constitute the hypernym of the term to be defined.

Finally, let us highlight some informative instances of the PCG and PCC hybrid subtrees that are considered in our approach: (1) non-adjacent subject-predicate patterns: $(NN, SBJ) \leftarrow (VBZ) \rightarrow (NN, PRD)$; (2) description of the definiens' head: $(NN, PRD) \rightarrow (IN) \rightarrow (NN, PMOD)$; and (3) synonymy relation between the genus and the hypernym in a definitional sentence: $(NN, PRD) \rightarrow (CC, COORD) \rightarrow (NN)$.

4   Experimental Setup

After reviewing and justifying the potential of dependency-based syntax analysis for DE, we discuss the features implemented for modelling our system. These are developed after parsing the initial set of documents with a graph-based parser [2].

1. **Subtrees**: From all the available three-node subtree combinations, we extract the following information for each node: Surface form (SF), Part-of-Speech (POS), and Dependency Relation (DR). Each sentence is transformed to a feature vector of the 15 most frequent subtrees of each type. The six types used in this feature as well as examples are shown in Table 1. Additionally, let us highlight that the number of subtrees used as features is not arbitrary. It comes from a manual analysis of the frequency distribution of each type across the dataset. The definitional sentences in the Wikipedia corpus tend to have recurrent syntactic patterns, which provokes a long-tailed frequency distribution and thus a remarkable gap between systematic and idiosyncratic features. By keeping only the 15 most frequent subtrees we design a consistent feature set across the types while at the same time disregarding the long tail in each type (see Figure 2 for an illustrative example of three types of subtrees).

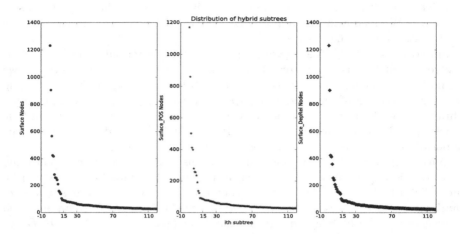

Fig. 2. Distribution of three selected types of hybrid subtrees illustrating the asymmetry of their frequency distribution

2. **Degree of X**: The *degree* of a node X in a graph is the number of edges adjacent to it, i.e. the sum of of its children + 1 (its head). We reduce the search space of X to

$$X \in \{PRD, SUBJ, APPO\}$$

because in this way, subject nodes with many modifiers are given more importance. For example, in the sample sentence in Figure 1, degree(PRD) = 4.

3. **Morphosyntactic chains starting in node X**: X can have the same node value as in the previous feature. If it exists, we extract all the children from that node in a recursive fashion. We then extract the POS and dependency relation chains and order it according to their index in the sentence. This approach has been used successfully in other NLP tasks, such as Semantic Role Labeling [10]. For example, in our sample sentence we would extract the following chain from the PRD node: $NMOD - PRD - COORD - CONJ - NMOD - PMOD..$, and so on untiil the last child is reached in recursive manner. Note that the head node (PRD) does not necessarily constitute the first element in the chain, as it has a child appearing before in the sentence ($(the, DT, NMOD)$).

4. **Ordered cartesian product of two subtrees**: The ordered cartesian product of two graphs G_1 and G_2 produces a new graph H with the vertex set $V(G_1) \times V(G_2)$, with the tuples $\{(i_1, i_2), (j_1, j_2)\}$ forming an edge if $\{i_1, j_1\}$ forms an edge in G_1 and $i_2 = j_2$, or $\{i_2, j_2\}$ forms an edge in G_2 and $i_2 = j_2$. Our intuition is that by extending the relationships between pairs of specific head nodes and their children, relations between modifiers of SBJ and PRD nodes, for example, would be captured and reinforced.

Table 1. Summary of the types of subtrees used and examples for each type. SF = Surface Form, POS = Part of Speech, DR = Dependency Relation.

Type of Subtree	Example
(SF, SF, SF)	(*TARGET*, refers, to) (is, used, in)
(POS, POS, POS)	(DT, JJ, NN) (IN, NN, VBD)
(DR, DR, DR)	(SBJ, ROOT, PRD) (PMOD, COORD, CONJ)
(SF + POS, SF + POS, SF + POS)	((is, VBZ), (a, DT), (unit, NN)), ((a, DT), (form, NN), (of, IN))
(SF + DR, SF + DR, SF + DR)	((In, LOC), (*TARGET*, SBJ), (was, ROOT)) ((is, ROOT), (any, PRD), (of, NMOD))
(POS + DR, POS + DR, POS + DR)	((NN, PMOD), (CC, COORD), (NN, CONJ)) ((DT, NMOD), (NNP, NAME), (NNP, PMOD))

We perform this operation only if the head of G_1 has the syntactic function SBJ and the head of G_2 has the head PRD or $APPO$. The result is a string that contains the surface, Part-of-Speech or dependency relation chain of the resulting graph H.

5. **Semantic similarity**: We hypothesize that we high semantic similarity between definiens and predicate, for example, might point towards a definitional sentence. In our sample sentence, this would be the case between *abortion* and *removal*. We extend this feature to other nodes like appositives or their postmodifiers, and apply it to the following pairs: (SBJ,PRD), (SBJ, $APPO$), (PRD, JJ_PMOD) and ($APPO$, JJ_PMOD). The Leacock Chodorow Similarity between each pair is computed, following the path between them in the WordNet hierarchy [14].

5   Evaluation

The above features are incorporated in a sentence' feature vector, and this information is used for training different classification algorithms present in the Weka workbench [37]. The evaluation results we report are based on 10-fold cross-validation. It is important to highlight that the Bag-of-Subtrees features are extracted only from the training set in each fold, which ensures there is no bias during the feature design stage, and the generalization potential is appropriately measured. Table 2 shows the scores for the different setups on which the experiments were carried out. For each algorithm, the different setups designed are: S_1 includes the full feature set, S_2 disregards chain and cartesian product features, and S_3 disregards chain, cartesian product, degree and similarity features. Likewise, Table 3 shows a comparative table among the different systems studied, which are the following:

- **Bigrams**: Baseline 1 based on the bigram classifier for soft pattern matching proposed by [5].

Table 2. Scores of our approach after being tested with several learning algorithms

	NaïveBayes			VPerceptron			SVM			LogisticR			DTrees			RandomF		
	P	R	F	P	R	F	P	R	F	P	R	F	P	R	F	P	R	F
S₁	81.9	78	75.9	78.9	85.2	84.4	**85.9**	**85.3**	**85.4**	83.4	82.7	84.7	85.7	85.3	85.2	83.4	82.9	82.2
S₂	75.7	75.9	75.5	80.5	79.4	79.5	82	81.2	81.3	82.2	81.5	81.6	80.8	79.4	79.6	79.9	79.4	79.5
S₃	53.1	58.6	49	56.9	59.8	52.2	56.9	59.8	52.2	56.9	59.8	52.2	55	59.2	49.3	56.9	59.8	52.2

Table 3. Comparative table of results between our approach and the reported scores in Navigli and Velardi (2010) and Yan et al. (2012)

	Our proposal	WCL 1	WCL 2	PMA	Star Patterns	Bigrams
Precision	85.9	**99.8**	99.8	89.1	86.7	66.7
Recall	85.3	42.1	60.7	**93.5**	66.1	82.7
F-Measure	85.4	59.2	83.5	**91.3**	75.1	73.9

- **Star Patterns**: Baseline 2 based on a pattern-matching approach in which infrequent words are replaced by '*' and then matched against candidate definitions [18].
- **WCL 1**: Implementation of the Word-Class Lattice model where a lattice is learned for each cluster of sentences [18].
- **WCL 3**: Implementation of the Word-Class Lattice model where a lattice is learned for each of the components of a definition, namely *Definiendum*, *Definitor* and *Definiens* [18].
- **PMA**: Pattern Mining Approach based on *Ngram*, subsequence and dependency subtrees patterns [38].
- **Our proposal**: An implementation of our approach with the highest scoring setup.

In the light of these scores, it seems reasonable to argue that a classification approach for DE can improve substantially by including features that account for the morphosyntactic structure of the sentence. Moreover, deep analyses of syntactic trees and the relation among dependents contributes decisively to DE. While syntactic and graph-based information is exploited in both [18] and [38], our results suggest that an approach exclusively reliant on these features can perform in a competitive fashion.

The highest scoring approach in terms of Precision is achieved by the WCL systems, with almost 1. The highest score in terms of Recall (83.5) and F-Measure (91.30), on the other hand, is the PMA system. After reviewing our own results, two main conclusions can be reached: (1) Our method improves substantially in terms of Recall and F-Measure with respect to the WCL approach, although it performs below the PMA approach in these metrics in about 10%[2]. (2) Although our approach does not outperform the aforementioned approaches, it does provide an insight towards highly informative features that could potentially be used in the future together with shallower features.

---

[2] An implementation of this method and an evaluation of the contribution of tree-based ngrams is expected to be carried out in the future.

6   Conclusions and Future Work

An approach to the task of DE based on morphosyntactic information derived from dependency relations has been described. From the initial hypothesis that syntactic features are highly discriminatory in predicting whether a sentence is or not a definition, we have justified the feature design and selection. Next, dataset on which the experiments have been carried out has been presented to the reader, followed by the experimental setup and the results, which seem to be encouraging enough to argue that features extracted from dependency relations can constitute the core of a successful DE system.

The main findings of this work can be summarized as follows:

- **Bag-of-Subtrees**: These features alone have proven to be uninformative enough as to constitute the backbone of a DE system. Drops in up to 30% in performance were identified after removing all the other features.
- **Subtree chain combinations**: Our experiments show that features that look at a subtree's children or the cartesian product of certain subtrees can contribute to improving Precision in a DE system. This is particularly significant in certain algorithms and setups, like Naïve Bayes, where an increase of 7% was achieved.

As for potential lines of future work, we are currently investigating four possible methods for improving the performance of DE systems. On one hand, we are interested in testing the extent to which semantic roles can help discriminating clusters of words as being part of a *definiens*, which we believe could be used as a previous step to DE. On the other hand, we are also looking at the potential of sequential labelling for DE. Previous work [11] [8] showed that this approach has several advantages over full-sentence classification. E.g. it allows identifying cross-sentence definitions, helps pruning out noise in very long sentences which encode short definitions, and helps identifying not only full definitions but also their components (e.g. definiens, definiendum or its genus, where present). Moreover, moving out of a domain like Wikipedia would constitue a real-world challenge for this approach and would be appropriate for testing its generalization potential. And finally, exhaustive error analysis would be advisable in order to understand the reasons behind some of the misclassified instances during our experiments. This information could be used for designing more efficient feature sets.

Acknowledgments. We would like to express our gratitude to the reviewers for their helpful comments. We also acknowledge partial support from project Dr. Inventor (FP7-ICT-2013.8.1 611383), programa Ramón y Cajal 2009 (RYC-2009-04291), and project number TIN2012-38584-C06-03 from Ministerio de Economía y Competitividad, Secretaría de Estado de Investigación, Desarrollo e Innovación, Spain.

References

1. Afzal, N., Mitkov, R., Farzindar, A.: Unsupervised relation extraction using dependency trees for automatic generation of multiple-choice questions. In: Butz, C., Lingras, P. (eds.) Canadian AI 2011. LNCS (LNAI), vol. 6657, pp. 32–43. Springer, Heidelberg (2011)
2. Bohnet, B.: Very high accuracy and fast dependency parsing is not a contradiction. In: Proceedings of the 23rd International Conference on Computational Linguistics, COLING 2010, pp. 89–97. Association for Computational Linguistics, Stroudsburg (2010)
3. Bontas, E.P., Mochol, M.: Towards a cost estimation model for ontology engineering. In: Eckstein, R., Tolksdorf, R. (eds.) Berliner XML Tage, pp. 153–160 (2005)
4. Borg, C., Rosner, M., Pace, G.: Evolutionary algorithms for definition extraction. In: Proceedings of the 1st Workshop in Definition Extraction (2009)
5. Cui, H., Kan, M.Y., Chua, T.S.: Generic soft pattern models for definitional question answering. In: Proceedings of the 28th Annual International ACM SIGIR Conference on Research and Development in Information Retrieval, pp. 384–391. ACM (2005)
6. Degórski, L., Marcińczuk, M., Przepiórkowski, A.: Definition extraction using a sequential combination of baseline grammars and machine learning classifiers. In: Proceedings of the Sixth International Conference on Language Resources and Evaluation, LREC 2008. ELRA, Marrakech (2008)
7. Del Gaudio, R., Batista, G., Branco, A.: Coping with highly imbalanced datasets: A case study with definition extraction in a multilingual setting. In: Natural Language Engineering, pp. 1–33 (2013)
8. Espinosa, L.: Towards definition extraction using conditional random fields. In: Proceedings of RANLP 2013 Student Research Workshop, pp. 63–70 (2013)
9. Fahmi, I., Bouma, G.: Learning to identify definitions using syntactic features. In: Proceedings of the Workshop on Learning Structured Information in Natural Language Applications, pp. 64–71 (2006)
10. Hacioglu, K.: Semantic role labeling using dependency trees. In: International Conference on Compcutional Linguistics, COLING (2004)
11. Jin, Y., Kan, M.Y., Ng, J.P., He, X.: Mining scientific terms and their definitions: A study of the ACL anthology. In: Proceedings of the 2013 Conference on Empirical Methods in Natural Language Processing, pp. 780–790. Association for Computational Linguistics, Seattle (2013)
12. Malaisé, V., Zweigenbaum, P., Bachimont, B.: Detecting semantic relations between terms in definitions. In: Ananadiou, S., Zweigenbaum, P. (eds.) International Conference on Computational Linguistics (COLING 2004) - CompuTerm 2004: 3rd International Workshop on Computational Terminology, Geneva, Switzerland, pp. 55–62 (2004)
13. Meyer, I.: Extracting knowledge-rich contexts for terminography. Recent Advances in Computational Terminology 2, 279 (2001)
14. Miller, G.A.: Wordnet: A lexical database for english. Communications of the ACM 38(11), 39–41 (1995)
15. Monachesi, P., Westerhout, E.: What can NLP techniques do for eLearning? In: International Conference on Informatics and Systems (INFOS 2008), pp. 150–156 (2008)
16. Muresan, A., Klavans, J.: A method for automatically building and evaluating dictionary resources. In: Proceedings of the Language Resources and Evaluation Conference, LREC (2002)

17. Nakamura, J.I., Nagao, M.: Extraction of semantic information from an ordinary english dictionary and its evaluation. In: Proceedings of the 12th Conference on Computational Linguistics, COLING 1988, vol. 2, pp. 459–464. Association for Computational Linguistics, Stroudsburg (1988)

18. Navigli, R., Velardi, P.: Learning word-class lattices for definition and hypernym extraction. In: Proceedings of the 48th Annual Meeting of the Association for Computational Linguistics, ACL 2010, pp. 1318–1327. Association for Computational Linguistics, Stroudsburg (2010)

19. Navigli, R., Velardi, P., Ruiz-Martínez, J.M.: An annotated dataset for extracting definitions and hypernyms from the web. In: Chair, N.C.C., Choukri, K., Maegaard, B., Mariani, J., Odijk, J., Piperidis, S., Rosner, M., Tapias, D. (eds.) Proceedings of the Seventh International Conference on Language Resources and Evaluation (LREC 2010), pp. 3716–3722. European Language Resources Association (ELRA), Valletta (2010)

20. Nivre, J.: Dependency grammar and dependency parsing. Tech. rep., Växjö University (2005)

21. Park, Y., Byrd, R.J., Boguraev, B.K.: Automatic Glossary Extraction: Beyond Terminology Identification. In: Proceedings of the 19th International Conference on Computational Linguistics, pp. 1–7. Association for Computational Linguistics, Morristown (2002)

22. Przepiórkowski, A., Spousta, M., Simov, K., Osenova, P., Lemnitzer, L., Kubo, V., Wójtowicz, B.: Towards the automatic extraction of definitions in Slavic. In: Proceedings ofo the BSNLP Workshop at ACL 2007, pp. 43–50 (2007)

23. Rebeyrolle, J., Tanguy, L.: Repérage automatique de structures linguistiques en corpus: le cas des énoncés définitoires. Cahiers de Grammaire 25, 153–174 (2000)

24. Reiplinger, M., Schäfer, U., Wolska, M.: Extracting glossary sentences from scholarly articles: A comparative evaluation of pattern bootstrapping and deep analysis. In: Proceedings of the ACL 2012 Special Workshop on Rediscovering 50 Years of Discoveries, pp. 55–65. Association for Computational Linguistics, Jeju Island (2012)

25. Saggion, H., Gaizauskas, R.: Mining on-line sources for definition knowledge. In: 17th FLAIRS, Miami Bearch, Florida (2004)

26. Sánchez, A., Márquez, J.: Hacia un sistema de extracción de definiciones en textos jurídicos. In: Actas de la 1er Jornada Venezolana de Investigación en Lingüística e Informática, pp. 1–10 (2005)

27. Sarmento, L., Maia, B., Santos, D., Pinto, A., Cabral, L.: Corpógrafo V3 From Terminological Aid to Semi-automatic Knowledge Engineering. In: 5th International Conference on Language Resources and Evaluation (LREC 2006), Geneva (2006)

28. Seppälä, S.: A proposal for a framework to evaluate feature relevance for terminographic definitions. In: Proceedings of the 1st Workshop on Definition Extraction, WDE 2009, pp. 47–53. Association for Computational Linguistics, Stroudsburg (2009)

29. Sierra, G., Alarcón, R., Aguilar, C., Barrón, A.: Towards the building of a corpus of definitional contexts. In: Proceeding of the 12th EURALEX International Congress, Torino, Italy, pp. 229–240 (2006)

30. Stevenson, M., Greenwood, M.A.: Comparing information extraction pattern models. In: Proceedings of the Workshop on Information Extraction Beyond The Document, IEBeyondDoc 2006, pp. 12–19. Association for Computational Linguistics, Stroudsburg (2006)

31. Storrer, A., Wellinghoff, S.: Automated detection and annotation of term definitions in German text corpora. In: Conference on Language Resources and Evaluation (LREC), pp. 275–295 (2006)
32. Sudo, K., Sekine, S., Grishman, R.: An improved extraction pattern representation model for automatic ie pattern acquisition. In: Proceedings of the 41st Annual Meeting of the Association for Computational Linguistics, ACL 2003, Sapporo, Japan (2003)
33. Szpektor, I., Tanev, H., Dagan, I., Coppola, B.: Barcelona
34. Walter, S., Pinkal, M.: Automatic extraction of definitions from German court decisions. In: Proceedings of the Workshop on Information Extraction Beyond the Document, pp. 20–28. Association for Computational Linguistics (2006)
35. Westerhout, E., Monachesi, P.: Extraction of Dutch definitory contexts for elearning purposes. In: Proceedings of the Computational Linguistics in the Netherlands (CLIN 2007), Nijmegen, Netherlands, pp. 219–234 (2007)
36. Westerhout, E., Monachesi, P.: Creating glossaries using pattern-based and machine learning techniques. In: Proceedings of the 6th Conference on Language Resources and Evaluation (LREC), pp. 3074–3081 (2008)
37. Witten, I.H., Frank, E.: Data Mining: Practical machine learning tools and techniques. Morgan Kaufmann (2005)
38. Yan, Y., Hashimoto, C., Torisawa, K.: Pattern mining approach to unsupervised definition extraction. In: Speech Processing Society 18th Annual Meeting (2005) (in Chinese)
39. Zhang, C., Jiang, P.: Automatic extraction of definitions. In: 2nd IEEE International Conference on Computer Science and Information Technology, pp. 364–368. IEEE (2009)

Entity Linking for Open Information Extraction

Arnab Dutta and Michael Schuhmacher

Research Group Data and Web Science, University of Mannheim, Germany
{arnab,michael}@informatik.uni-mannheim.de

Abstract. Open domain information extraction (OIE) projects like NELL or REVERB are often impaired by a schema-poor structure. This severely limits their application domain in spite of having web-scale coverage. In this work we try to disambiguate an OIE fact by referring its terms to unique instances from a structured knowledge base, DBPEDIA in our case. We propose a method which exploits the frequency information and the semantic relatedness of all probable candidate pairs. We show that our combined linking method outperforms a strong baseline.

1 Introduction

In the recent past, there have been major developments in the area of open domain information extraction (OIE). Projects like NELL [2] or REVERB [5] have introduced an era of *open* information extraction systems which are characterized by their web-scale coverage at the expense of a poor schema. On the other extreme, Wikipedia based extraction systems like DBPEDIA [1] or YAGO [11] provide more structured information but have limited coverage.

The data maintained by OIE systems is important for analyzing, reasoning about, and discovering novel facts on the web and has the potential to result in a new generation of web search engines [4]. But the lack of a proper schema severely limits the applicability of such data. Moreover, facts from the OIE are often too ambiguous. For instance, a typical fact extracted by NELL might be *bookwriter(imperialism, lenin)*. While we might have an intuitive understanding of the property *bookwriter*, it is difficult to determine the correct references of the subject and object terms within the triple. Here, the surface form object *lenin* can refer to Vladimir Lenin (the Russian political theorist), Lenin (a nuclear icebreaker) or the Lenin Prize. For that reason, we need to disambiguate the NELL terms to uniquely identifiable instances. We opt to use DBPEDIA as the structured knowledge-base providing globally unique URIs for each instance.

In general, the entity linking task is to match surface form mentions from a natural language text to the corresponding knowledge base instances. However, in this work, we focus on linking ambiguous NELL subjects and objects to DBPEDIA. While established Entity Linking systems, like e.g. DBPEDIA Spotlight [8] or Aida [7], exploit, besides other features, mainly the context of the entity within the text, in our case of OIE triple linking, this context information is in many cases not available or does not exist. This sets our problem setting apart from the traditional linking task.

E. Métais, M. Roche, and M. Teisseire (Eds.): NLDB 2014, LNCS 8455, pp. 75–80, 2014.

2   Entity Linking Methods

In the following, we present first a simple, yet very strong, baseline method, namely the Frequency-based Entity Linking as used in [3]. Second, we introduce a knowledge-based approach which exploits the DBPEDIA ontology itself, following the DBPEDIA graph exploration method introduced by [10]. Last, we propose a Combined Entity Linking approach, which incorporates the frequency-based with the graph-based approach. Note that in this work, we do not focus on mapping predicates from NELL to corresponding DBPEDIA properties because (a) existence of corresponding DBPEDIA properties cannot be guaranteed and (b) an exact analogous mapping may not be possible, for instance the NELL property *agentcollaborateswithagent*, could be mapped to any one of dbp:influences, dbo:publisher or dbo:employer.

Frequency-Based Entity Linking. A simple, yet high performing approach for mapping a given surface form (NELL subject/objects in our case), to its corresponding DBPEDIA (or Wikipedia) instance is to link to its most frequent candidate entity [9]. Even though this approach does not take any context information into account, it has been proven to be effective not only for text entity linking, but also for NELL triple linking [3]. We thus use it as a baseline method. For obtaining the frequencies, the intra-Wikipedia links connecting anchor texts to article pages were exploited (using WikiPrep [6]). The assumption was that, the maximum number of outgoing links from an anchor to a particular article marks the article as the most probable entity referred to by its surface form (the anchor). Formally, if an anchor e refers to N Wikipedia articles A_1, ..., A_N with n_1, ..., n_N respective link counts, then the conditional probability P of e referring to A_j is given by, $P(A_j|e) = n_j / \sum_{i=1}^{N} n_i$. Thus, the pair (e, A_j), henceforth called subject/object-instance-mapping, is awarded the probability P. For every NELL triple, we can use this approach since each DBPEDIA entity is equal to its Wikipedia article title. We rank the candidates on descending P and define a *top* ranked list as $E_{Subj|top-k}$ (for subject mappings) and $E_{Obj|top-k}$ (for object mappings). Since every mapping of a subject is independent of the object mapping, we compute the prior probability as $P_{prior} = P_{Subj}P_{Obj}$. We select the DBPEDIA subject-object pair with the highest prior probability.

Graph-Based Entity Linking. We assume that the subject and object connected by a NELL predicate are related to each other and that some relationship can be found within the DBPEDIA knowledge base. This motivates us to exploit the latent contextual connectivity between NELL terms instead of relying just on the most frequent entity.

We obtain the likelihood of each possible pair of subject-object candidates by computing the semantic relatedness [12] between subject and object. As we do not want to make any assumptions about the existence or the type of the DBPEDIA properties to be taken into account, we adapt the property-agnostic approach presented in [10] and summarized as below.

1) We consider all combinations $E_{Subj|top-5} \times E_{Obj|top-5}$ and compute each pairwise cheapest path, treating DBPEDIA as a semantic network (see [10] for details).

2) We weigh the DBPEDIA graph edges, thus automatically capturing the importance of different property edges, by an information-content- based measure (CombIC) which was reported as the best of the graph-weighting schema proposed by [10] for computing semantic similarity.

3) We select the subject-object pair from $E_{Subj} \times E_{Obj}$ which has the minimum path cost, on the weighted graph. The path cost between two entities is calculated as the sum of the edge costs along their undirected connecting path and is normalized as probabilities to P_{graph}

As result, we jointly disambiguate subject and object to their most semantically similar DBPEDIA candidate entities.

Combined Entity Linking. Our last approach is motivated by the fact that both approaches the frequency-based and the graph-based linking have an individual weakness, but can complement each other. The former exploits the empirically obtained frequency data about common surface-form-to-instance mappings, however, it cannot incorporated the information that subject and object should most likely be related somehow. But this information is used by the graph-based linking, which finds this vague relationship between subject and object in the background knowledge base DBPEDIA, however, ignoring the important frequency information. Consequently, we opt for a linear combination of the two approaches and select the subject-object combination with the highest combined probability

$$P_{comb} = \lambda P_{graph} + (1 - \lambda) P_{prior}$$

where the weighting factor λ is set to 0.5 initially, thus giving equal influence to the graph and the frequency information. With this combination, we give preference to those subject-object combinations, having individually high likelihoods and which are also closely semantically related in the DBPEDIA knowledge-base.

3   Experiments

Dataset and Metric. We use the gold standard from [3][1], which consists if 12 different NELL properties with 100 triples each that have been manually linked to their correct DBPEDIA entities. For our evaluation, we excluded the predicate *companyalsoknownas*, as it contains actually not distinct subjects and object, but only different surfaces forms for the same entity, e.g. *companyalsoknownas(General Motors, GM)*. As metric, we use Precision (P), Recall (R) and F-measure (F_1), and evaluate each subject and object mapping individually, thus also accounting for partially correct triple.

---

[1] Downloaded from `https://madata.bib.uni-mannheim.de/65/`

Table 1. Performance scores of our proposed methods and the baseline. Best F_1 values for each predicate is marked in bold.

	Frequency-based			Graph-based			Combined		
	P	R	F_1	P	R	F_1	P	R	F_1
actorstarredinmovie	80.7	82.0	81.3	89.8	91.2	90.5	91.4	92.8	**92.1**
agentcollaborateswithagent	81.6	85.9	**83.7**	69.3	72.9	71.1	81.6	85.9	**83.7**
animalistypeofanimal	85.7	88.0	**86.8**	62.4	64.1	63.3	85.2	87.5	86.3
athleteledsportsteam	88.6	85.5	87.0	87.0	84.0	85.5	91.7	88.5	**90.1**
bankbankincountry	81.7	77.6	**79.6**	68.3	64.8	66.5	81.7	77.6	**79.6**
citylocatedinstate	79.0	79.4	79.2	81.5	81.9	81.7	86.0	86.4	**86.2**
bookwriter	82.2	83.1	82.6	83.8	84.7	84.2	87.6	88.5	**88.0**
personleadsorganization	83.6	79.0	81.2	78.4	74.0	76.1	84.8	80.1	**82.4**
teamplaysagainstteam	81.8	81.8	81.8	61.0	61.0	61.0	85.6	85.6	**85.6**
weaponmadeincountry	88.9	87.0	**87.9**	44.4	43.5	44.0	84.7	82.9	83.8
lakeinstate	90.3	93.0	91.6	84.7	86.6	85.6	91.5	93.6	**92.5**
Average all 11 predicates	84.0	83.8	83.9	73.7	73.5	73.6	86.5	86.3	**86.4**

Results and Analysis. We report the performance for each of the three methods in Table 1; the frequency-based, the graph-based, and the combined approach. As expected, we find that the most frequent entity baseline shows strong results. In contrast, the graph-based method shows an overall F_1-measure of only 73.6 compared to 83.9 for the baseline. Our combining approach, however, improves over the most frequent entity baseline by 2.9% w.r.t. average F_1, which is notably a difficult competitor for unsupervised and knowledge-rich methods.

When analyzing the results in detail, we find the combined approach to improve the F_1-measure for all but two NELL predicates. In contrast, the graph-based approach, which does not take into account any information about the term interpretation frequencies, has a great variation in performance: for instance in *actorstarredinmovie*, F_1 increases from 81.3 to 90.5, but for *weaponmadeincountry*, it decreases by 50%, the latter meaning that the graph-based method selects very often highly related, but incorrect subject-object pairs. Analyzing this different performances more detailed, we attribute the improvement to the fact that the underlying knowledge base had sufficient relatedness evidence favoring the likelihood of the correct candidate pairs. For example for *actorstarredinmovie*(morgan freeman, seven), two possible candidate pairs (out of many others) with their probabilities are as follows:

$(\texttt{dbp:Morgan\_Freeman}, \texttt{dbp:Seven\_Network})$ $P_{prior} = 0.227;$ $P_{graph} = 0.074$

$(\texttt{dbp:Morgan\_Freeman}, \texttt{dbp:Seven\_(film)})$ $P_{prior} = 0.172;$ $P_{graph} = 0.726$

With the most frequent entity method, we would have selected the former pair, given its higher prior probability of $P_{prior} = 0.227$. However, the graph-based method captures the relatedness, as DBPEDIA contains the directly connecting edge $\texttt{dbo:starring}$ and thus rightly selects the later pair. In other cases, as observed often with *personleadsorganization* and *weaponmadeincountry*, a low prior probability was complemented with a semantic relatedness, thus a high P_{graph},

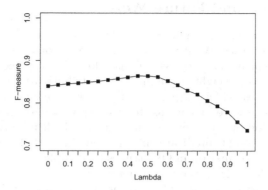

Fig. 1. Effect of Lambda (λ) on the average F_1 score

thereby making a highly related, but incorrect subject-object-combination candidate more likely than the correct one. Consequently, the graph-based approach by itself lowers the performances, relative to the baseline.

The fact that our combined method outperforms both the other approaches indicates that the linear combination of the two probabilities effectively yields in selecting the better of the two methods for each NELL triple. However, in addition to this effect, we observe that our combined approach also finds the correct mapping in cases where both, the frequency-based and the graph-based approach fail individually. Giving one example from the data, for the triple *teamplaysagainstteam(hornets, minnesota timberwolves)*[2], the frequency-based approach disambiguates it to the pair (`dbp:Hornet, dbp:Minnesota_Timberwolves`), which is incorrect, as `dbp:Hornet` is an insect. But the graph-based approach also disambiguates wrongly to the pair (`dbp:Kalamazoo_College, Minnesota_Timberwolves`), even though it discovers a very specific path in DBPEDIA between subject and object in this pair, via the intermediate entity `dbp:David_Kahn_(sports_executive)`. The gold standard pair, (`dbp:New_Orleans_Pelicans, dbp:Minnesota_Timberwolves`), however, gets selected by the combined approach, which combines the medium high prior probability and a medium high relatedness originating from the fact that both instances are connected by `yago:YagoLegalActor`. Not that this last information originates from DBPEDIA and its unsupervised graph weighing method, not from the NELL predicate *teamplaysagainstteam*.

Last, we report on the robustness of our combined approach with respect to the parameter λ, even though giving equal weight to both methods, thus setting λ to 0.5, seems to be a natural choice. Figure 1 shows the F_1-measure for $\lambda \in [0; 1]$. Note that $P_{joint} = P_{graph}$, when $\lambda = 1$ and $P_{joint} = P_{prior}$, when $\lambda = 0$. We observe a clear peak at $\lambda = 0.5$, which confirms our initial choice.

---

[2] "hornets" refers to `dbp:New_Orleans_Pelicans`, formerly the New Orleans Hornets.

4   Conclusions and Future Work

We addressed the linking of ambiguous NELL subject and object terms to their unique DBPEDIA entities. We studied the effectiveness of a simple baseline which uses the most frequent instance, as well as a knowledge-based approach which exploits DBpedia as a weighted graph. Our contribution is the combination of these two approaches, which outperforms the individual methods at a high level of 86.4 for the F_1-measure. In contrast to other approaches, our method does not require any learning or parameter tuning and the high performance is achieve without using any NELL predicate information. Essentially, we overcome the lack of contextual information in OIE triples by complementing it with existing background knowledge from the target ontology.

As part of future work, we will incorporate the property information from NELL to improve the entity disambiguation. Ultimately, we aim for a NELL predicate disambiguation to DBPEDIA that resolves complex one-to-many property mappings, which could be extracted from those paths currently selected within the cheapest path computation of the graph-based entity linking approach.

References

1. Bizer, C., Lehmann, J., Kobilarov, G., Auer, S., Becker, C., Cyganiak, R., Hellmann, S.: DBpedia – A Crystallization Point for the Web of Data. Journal of Web Semantics 7(3) (2009)
2. Carlson, A., Betteridge, J., Kisiel, B., Settles, B., Hruschka Jr., E.R., Mitchell, T.: Toward an architecture for never-ending language learning. In: Proc. of AAAI 2010 (2010)
3. Dutta, A., Niepert, M., Meilicke, C., Ponzetto, S.P.: Integrating open and closed information extraction: Challenges and first steps. In: Proc. of the ISWC 2013 NLP and DBpedia Workshop (2013)
4. Etzioni, O.: Search needs a shake-up. Nature 476(7358), 25–26 (2011)
5. Etzioni, O., Cafarella, M., Downey, D., Kok, S., Popescu, A.-M., Shaked, T., Soderland, S., Weld, D.S., Yates, A.: Web-scale information extraction in KnowItAll (Preliminary results). In: Proc. of WWW (2004)
6. Gabrilovich, E., Markovitch, S.: Overcoming the brittleness bottleneck using wikipedia: Enhancing text categorization with encyclopedic knowledge. In: Proc. of AAAI 2006 (2006)
7. Hoffart, J., Yosef, M.A., Bordino, I., Fürstenau, H., Pinkal, M., Spaniol, M., Taneva, B., Thater, S., Weikum, G.: Robust disambiguation of named entities in text. In: Proc. of EMNLP 2011 (2011)
8. Mendes, P.N., Jakob, M., García-Silva, A., Bizer, C.: DBpedia Spotlight: Shedding light on the web of documents. In: Proc. of I-Semantics (2011)
9. Mihalcea, R., Csomai, A.: Wikify! Linking documents to encyclopedic knowledge. In: Proc. of CIKM 2007 (2007)
10. Schuhmacher, M., Ponzetto, S.P.: Knowledge-based graph document modeling. In: Proc. of WSDM 2014 (2014)
11. Suchanek, F.M., Kasneci, G., Weikum, G.: YAGO: A core of semantic knowledge. In: Proc. of WWW 2007 (2007)
12. Zhang, Z., Gentile, A.L., Ciravegna, F.: Recent advances in methods of lexical semantic relatedness – a survey. Natural Language Engineering 1(1), 1–69 (2012)

A New Method of Extracting Structured Meanings from Natural Language Texts and Its Application

Vladimir A. Fomichov and Alexander A. Razorenov

Dept of Innovations and Business in the Sphere of Informational Technologies,
Faculty of Business Informatics, National Research University Higher School of Economics,
Kirpichnaya str. 33, 105187 Moscow, Russia
vfomichov@hse.ru, vfomichov@gmail.com

Abstract. An original method of developing the algorithms of semantic-syntactic analysis of texts in natural language (NL) is set forth. It expands the method proposed by V.A. Fomichov in the monograph published by Springer in 2010. For building semantic representations, the class of SK-languages is used. The input texts may be at least from broad and practically interesting sub-languages of English, French, German, and Russian languages. The final part of the paper describes an application of the elaborated method to the design of a NL-interface to an action-based software system. The developed NL-interface NLC-1 (Natural Language Commander -Version 1) is implemented with the help of the functional programming language Haskell.

Keywords: natural language processing, algorithm of semantic-syntactic analysis, theory of K-representations, SK-languages, semantic representation, text meaning representation, action-based application, Natural Language Commander, Haskell.

1   Introduction

The natural language interfaces (NL-interfaces) to databases and knowledge bases are to extract the meaning of a query and to reflect it in a problem-oriented expression. Most often, the subsystems of NL-interfaces transforming input texts into the representations of their meanings use a form of syntactic representation as an intermediate level. The most popular choices for syntactic parser are the Stanford parser and Charniak parser. The NL-interface to knowledge bases ORAKEL [1] includes a syntactic Early-type parser. In [4], a developed Montague-grammar-based syntactic analyzer is described. It is a part of the NL-interface transforming the queries in NL into the expressions of the Web search language SPARQL.

However, if we consider the question what is the essence of mastering mother tongue by the child, most likely, we conclude that the young child has no idea about syntactic categories, and the understanding of language is based on the association of words with different meanings, on the knowledge of various morphological forms of the words, and on taking into account the order of words. This observation enables us to conjecture that it is possible to find a way of constructing a text's semantic

E. Métais, M. Roche, and M. Teisseire (Eds.): NLDB 2014, LNCS 8455, pp. 81–84, 2014.

representation (SR), not using a pure syntactic representation as an intermediate level but in a more straightforward manner, proceeding from the meanings of lexical items, the values of their morphological properties, and the order of lexical items.

This conjecture was proved in the monograph [2], introducing the current configuration of the theory of K-representations (knowledge representations). It includes, in particular, the definition of a new class of formal languages called SK-languages (standard knowledge languages). Besides, the monograph introduces an original method of transforming NL-texts into their semantic representations. The steps of transformation don't include the construction of a syntactic representation. Instead of this step, an original form of semantic-syntactic presentation is used, it is a special string-digital matrix called a matrix semantic-syntactic representation of the input text. The method stated in [2] was realized in the same book as a multilingual algorithm of semantic-syntactic analysis called *SemSynt1*.

The main subject of this paper is to set forth an expanded version of the method of transforming natural language texts into semantic representations introduced in [2]. The second subject of the paper is a description of the method's application to the design of a NL-interface to an applied computer system, it makes easier for the non-programming users the interaction with the file system of a computer.

2   Task Statement

The monograph [2] introduces an original method of transforming NL-texts into SRs: (a) Phase 1. A component-morphological analysis of the input text; (b) Phase 2. The construction of a matrix semantic-syntactic representation of the text (MSSR); (3) Phase 3. The assembly of semantic representation of the input text being its K-representation (KR), i.e., an expression of a certain SK-language.

The main assumptions of the method: (1) The input texts may be the questions of many kinds, the commands, the sentences, and the discourses; (2) The input texts don't include the logical connective "or", participle constructions, gerundial constructions, and subordinate clauses; (3) The texts don't include the phrases describing the value of a certain function for the mentioned arguments; (4) No phrases describing the relationships between the entities with the help of comparative attributes.

The first task of this paper is to expand the method introduced in [2], taking into account all NL phenomena listed above in the items (1) – (4). The second task of the paper is to outline the central ideas of using this method for the design of a NL-interface to an applied computer system helping the end users to work with the file system of a computer.

3   The Proposed Modifications of a Method of Extracting the Meanings from Natural Language Texts

1. Many notions corresponding to the words and word combinations from NL-texts are too general in order to be used for the interaction with a database. For instance, these are the concepts "IT-specialist" and "alumni". That is why it is proposed to use for semantic-syntactic processing of NL-texts not only a linguistic database but also a linguistic knowledge base. It may consist of the K-strings of the form illustrated by the following example: *(IT-specialist ≡ person * (Qualification, (programmer OR database-administrator OR web-programmer)))*. Let's call *unfolding concepts* the concepts being the left parts of some expressions in the linguistic knowledge base.

2. It is proposed to add to the method described in [2] a new final step. Its content is to replace all semantic items from the constructed primary SR belonging to the sub-class of unfolding concepts by the less general concepts with the help of the definitions stored by the used linguistic knowledge base (it may be interpreted as a part of ontology). For instance, the concept "IT-specialist" will be replaced by the compound concept *person * (Qualification, (programmer OR database-administrator OR web-programmer))*.

3. It is proposed to add to an MSSR the final numerical column with the index *contr* (as last column). It is necessary for connecting the main clause and subordinate clauses and a noun with a participle construction.

Example. Let T1 = "The company "Rainbow" has opened an office in the city where its President lived in the 1990s". A marked version T1mrk (with the distinguished tokens) of this text is as follows: "The company (t1) "Rainbow" (t2) has opened (t3) an office (t4) in (t5) the city (t6) where (t7) its (t8) President (t9) lived (t10) in (t11) the 1990s (t12)". Then Matr [6, contr] = 7, Matr [7, contr] = 6.

4. Suppose that a phrase describes the value of a certain function for certain arguments ("The price of this car is 14,000 euros", etc.). Then the edges in the graph induced by the constructed MSSR Matr go from the name of the function to the designations of the arguments and are to have the marks *arg1, arg2*, etc. The edge from the name of the function to its value is to have the mark *value*.

5. Let a phrase describe a relationship between several entities. Then the edges in the graph induced by the constructed MSSR Matr go from the name of the relationship to the designations of the attributes and are to have the marks *attr1, attr2*, etc.

4   Application of the Method

An expanded method of extracting meanings from NL-texts was used as a framework for the development of the algorithm of semantic-syntactic analysis *SemSynt2*. Using *SemSynt2*, we have developed the first version of a Natural Language Commander (NLC-1) – a file manager with a NL-interface. The main goal of this project is to create a workable theory of *Natural Language Interfaces to Action-Based Applications* and a useful software system for human-computer interaction. This application constructs a K-representation (KR) from the user's input instruction and then transforms it into *unified form*. This unified form is also a KR where complex concepts like music, video, etc. are replaced by basic concepts. E.g., the concept "music" is interpreted as the files with the extensions "mp3", "ogg", "acc". The application transforms this unified KR into a script of operation system shell – now it's Bourne-again shell (bash) both for Unix-like OS and for Windows – and executes this script.

Let's look how NLC-1 processed the input "Copy music files from "Download" folder to folder with name "Music" or "My music" on backup drive if their size is less than 1 GB". This instruction has the following primary KR constructed by *SemSynt2*: *If-then(Less(SizeOf(all music1*(Place1, certain folder1 * (Name1, "Download")):o1), 1/GB), Command (#Operator#, #Executor#, #now#, Copying1 *(Source1, o1)(Destination1, certain folder1 * (Name1, ("Music" OR "My music"))(Place1, certain backup-drive))))*.

Now if knowledge base of NLC-1 contains the K-strings *(music1 ≡ file1 * (Extension, ("mp3" OR "ogg" OR "wav" OR "aac"))), (backup-drive ≡ drive1 *(Name1,"F"))* and knowledge management system includes the rule *(x, x ≡ y ⊢ y)*, then NLC-1 transforms the constructed primary K-representation of user instruction into its secondary KR *If-then(Less(SizeOf(all file1 *(Extension, ("mp3" OR "ogg" OR "wav" OR "aac"))(Place1, certain folder1 * (Name1, "Download")) : o1), 1/GB), Command((#Operator#, #Executor#, #now#, Copying1*(Source1,o1)(Destination1, certain folder1*(Name1, ("Music" OR "My music"))(Place1, certain drive1 * (Name1, "F")))))*.

Then the result shell script is *if [$(du -cb "Download/*.mp3" "Download/*.ogg" "Download/*.wav" "Download/*.acc"|grep total|sed -e "s/\s.*$//g") -le 1000000000]; then cp "Download/*.mp3" "Download/*.ogg" "Download/*.wav" "Download/*.acc" $(ls /f/|grep -iE "^Music$/^My music$" | head -n1); fi*

Written in Haskell, NLC-1 is a flexible and scalable application. It can be configured by a researcher for different domains and underlying shells.

5   Conclusion

The preconditions for the elaboration of a new method (no-pure-syntax-use method) of transforming NL-texts into SRs were created by high expressive power and flexibility of SK-languages. It seems that the expanded method introduced above can be useful for developing a Multilingual Semantic Web (see also [2, 3]).

References

1. Cimiano, P., Haase, P., Heizmann, J., Mantel, M.: ORAKEL: A Portable Natural Language Interface to Knowledge Bases. Technical Report. Institute AIFB, University of Karlsruhe, Germany (2007)
2. Fomichov, V.A.: Semantics-Oriented Natural Language Processing: Mathematical Models and Algorithms. Springer, Heidelberg (2010)
3. Fomichov, V.A.: A Broadly Applicable and Flexible Conceptual Metagrammar as a Basic Tool for Developing a Multilingual Semantic Web. In: Métais, E., Meziane, F., Saraee, M., Sugumaran, V., Vadera, S. (eds.) NLDB 2013. LNCS, vol. 7934, pp. 249–259. Springer, Heidelberg (2013)
4. Ferré, S.: SQUALL: A Controlled Natural Language as Expressive as SPARQL 1.1. In: Métais, E., Meziane, F., Saraee, M., Sugumaran, V., Vadera, S. (eds.) NLDB 2013. LNCS, vol. 7934, pp. 114–125. Springer, Heidelberg (2013)

A Survey of Multilingual Event Extraction from Text

Vera Danilova, Mikhail Alexandrov, and Xavier Blanco

Autonomous University of Barcelona,
Edificio B, Campus UAB 08193
Bellaterra (Cerdanyola del Vallés), Barcelona, Spain
maolve@gmail.com, MAlexandrov@mail.ru, Xavier.Blanco@uab.cat
http://www.uab.cat/letras/

Abstract. The ability to process multilingual texts is important for the event extraction systems, because it not only completes the picture of an event, but also improves the algorithm performance quality. The present paper is a partial overview of the systems that cover this functionality. We focus on language-specific event type identification methods. Obtaining and organizing this knowledge is important for our further experiments on mono- and multilingual detection of socio-political events.

Keywords: Event extraction systems, Event type detection, Multilinguality.

1 Introduction

Multilingual functionality expands information access borders for question answering, data retrieval, summarization, and other applications. Event extraction (EE) systems currently improve language coverage in order to monitor the emergence of specific events in target regions, where valuable data appears in local languages. Also, multilinguality contributes to other EE applications: person-oriented (e.g., personalized news systems, tourist recommender systems), science-oriented (e.g., sociological, linguistic studies), etc. Multilingual knowledge is important, because its provides complementarity and less biased view of events and opinions [10]. Also, cross-lingual information fusion is reported to improve the quality of event extraction performance [7].

The objective of this overview is to outline the mechanisms of language-specific event type detection in several multilingual event extraction systems (MEES) that will be taken into account in the implementation of our experiments (socio-political event detection in news streams). It should be mentioned that we do not pretend to cover the multilingual event extraction issue in its entirety, as we went into system details not long ago and are currently at the stage of pilot experiments. Also, due to the insufficiency and nonuniformity of data on system evaluation, we decided not to address this issue in this paper.

In Section 2, related work, as well as event understanding and representation, are discussed. Section 3 describes several MEES and their mechanisms for event type identification. Section 4 concludes the paper.

E. Métais, M. Roche, and M. Teisseire (Eds.): NLDB 2014, LNCS 8455, pp. 85–88, 2014.

2   Related Work and Event Understanding

An overview of monolingual event extraction approaches is given in [5]. They are classified as data-driven, knowledge-driven and hybrid, and are evaluated with respect to the amount of training data, expert knowledge and expertise required, as well as the interpretability of results. The paper covers only the period of 1992-2010, therefore, there is a need of further description of monolingual, as well as multilingual approaches and systems.

An event is a predication with arbitrary number of arguments and relationships [3,8]. Event-related information is commonly represented in the form of a report on *who did what to whom, where and when, through what methods and why* (Automated Content Extraction Program (ACE): http://www.itl.nist.gov/iad/mig/tests/ace/), with a varying number of components depending on the system task. The report is further stored as a database record (a scenario template with a pre-defined number of slots). According to ACE terminology, *event trigger* is the word that determines the event occurrence; *argument* is an entity mention, a value or a temporal expression that constitute event attributes and *event mention* is an extent of text with the distinguished trigger, entity mentions and other argument types [3].

3   MEES: Event Type Detection

Earlier MEES, such as MURASAKI (1994) [1], use machine translation for document representation (in English or another *interlingua*) and apply monolingual extraction methods. The main drawback of MT use is the quality of translation, e.g. of proper names or domain-specific terms. State-of-the-art approaches prefer filtering and processing text in the original language [10] with the further event merging in a database. The MEES general pipeline includes data crawling and gathering, document pre-processing, event type detection, slot filling, database population and information fusion (cross-article and cross-language).

This article covers only four MEES due to space constraints: Xlike project (xlike.org), PULS (puls.cs.helsinki.fi/static/index.html), ZENON [4], and NEXUS [7].

Xlike (2012-2014) uses Newsfeed (newsfeed.ijs.si) service as data source. It processes news articles and blog entries in English, German, Spanish, Chinese, French, Italian, Portuguese and Dutch in order to mine events of any type, categorize them using a taxonomy and store them to the global event registry.

PULS (2006-present) is a project of the University of Helsinki. It uses European Media Monitor (EMM) MedISys News feed as the main data source. The system assists the analysts in monitoring epidemic threats, cross-border crime and illegal migration at the EU borders. There are separate modules for three languages: English, French and Russian.

ZENON (2001-2011) assists the experts in intelligence reports analysis. It covers conflicts (between ethnic groups, political parties), infrastructure problems and person-related events (marriage, meeting, etc.) in the operational area of

Table 1. MEES: sources and methods for event type detection

| Project | Xlike | PULS | | | ZENON | EMM-NEXUS |
		English	French	Russian		
Knowledge Base	Taxonomy (dmoz.org)	Scenario-specific ontologies	-	Modified ontologies for epidemic and security domains	Domain- and application-specific ontology	Event type hierarchy
Language Resources	Aligned multilingual corpus, GeoNames DB (geonames.org)	Scenario-specific lexica	Location DB(400 terms), Disease DB (150 terms)	Disease dictionary, shared lexicon, GeoNames DB	Handcrafted full-form lexica, gazetteers (Dari, Tajik)	Domain-specific and domain-independent dictionaries, gazetteers
Event structure	Title, time, location, main NE	Disease, time, location, status, cases (people involved), description	Disease name, location, cases, date	Disease, time, location, status, cases, description	FrameNet representation of situation type (action type + semantic roles)	Event type, date, location, number of killed, injured, kidnapped etc.
Event type detection method	Clustering (hierarchical bisecting *k*-means)	Patterns-matching (regex trigger+action code)	Discourse-based rules application	Pattern-matching (regex trigger+action code)	Verb Phrase Transducers, Semantic frames filling	Clustering, pattern-matching (1-2 slot: semantic roles + string attribute)
Pattern acquisition method	-	Semi-supervised bootstrapping	-	Manual	-	Weakly supervised bootstrapping

the German Armed Forces in Afghanistan. The system processes English, Dari and Tajik texts from the KFOR Corpus [4].

NEXUS is a software developed for the EMM (1999-present), a project of the Joint Research Center of the European Commission. EE engine is developed in collaboration with PULS. It processes events, related to illegal migration, cross-border crime, and crisis situations at and beyond EU borders. Active language coverage is as follows: English, Italian, Spanish, French, Portuguese, Russian, Arabic. Clustered source data is supplied by the EMM news aggregator.

Sources and methods for event type detection are indicated in Table 1. Xlike and NEXUS both use a cluster-centric approach as opposed to PULS and ZENON. Some of the systems process the entire article text (Xlike, ZENON, English and Russian modules of PULS) and the others - only the title and a few sentences (NEXUS, French PULS). There is a lack of event relevance ranking for the user in ZENON and Xlike systems. Xlike only measures event relevance to a certain taxonomy category. In event type detection, PULS needs to query ontologies to associate the trigger with a class, while NEXUS patterns are independent of the knowledge base. French PULS approach is the most robust as it uses only discourse-specific rules in combination with minimum disease and location lexica. It yields competitive results. In ZENON verb phrase transducers are used to detect verb phrases and subsequent action types. When an action is identified, the corresponding semantic frame (FrameNet (framenet.icsi.berkeley.edu/fndrupal/)) is filled.

4   Conclusion

This paper is a partial survey of MEES. A detailed examination and analysis of the present projects, as well as of several other systems (Biographe (biographe.org), KYOTO (kyoto-project.eu), NetOwl (netowl.com), Newstin (newstin.com), etc.) and multilingual EE frameworks, such as SProUT platform (sprout.dfki.de), GATE (gate.ac.uk), will be described in an upcoming paper.

References

1. Aone, C., Blejer, H., Okurowski, M.E., Van Ess-Dykema, C.: A Hybrid Approach to Multilingual Text Processing: Information Extraction and Machine Translation. In: Proceedings of the First Conference of the AMTA (1994)
2. Atkinson, M., Piskorski, J., Van der Goot, E., Yangarber, Y.: Multilingual Real-time Event Extraction for Border Security Intelligence Gathering. In: Wiil, U.K. (ed.) Counterterrorism and Open Source Intelligence. Lecture Notes in Social Networks 2, pp. 355–390. Springer, Wien (2011)
3. Grishman, R.: Information Extraction: Capabilities and Challenges. In: Notes for the 2012 International Winter School in Language and Speech Technologies. Rovira i Virgili University, Tarragona (2012)
4. Hecking, M., Sarmina-Baneviciene, T.: A Tajik extension of the multilingual information extraction system ZENON. In: CCRP C4ISR Cooperative Research Program, Washington, DC, p. 14 (2010)
5. Hogenboom, F., Frasincar, F., Kaymak, U., de Jong, F.: An Overview of Event Extraction from Text. In: CEUR Workshop Proceedings, vol. 779, pp. 48–57 (2011)
6. Lejeune, G.: Structure patterns in Information Extraction: a multilingual solution? In: Advances in Method of Information and Communication Technology AMICT 2009, Petrozavodsk, Russia, vol. 11, pp. 105–111 (2009)
7. Piskorski, J., Belayeva, J., Atkinson, M.: Exploring the Usefulness of Cross-lingual Information Fusion for Refining Real-time News Event Extraction: A Preliminary Study. In: RANLP 2011, pp. 210–217 (2011)
8. Piskorski, J., Yangarber, R.: Information Extraction: Past, Present and Future. In: Poibeau, T., et al. (eds.) Multi-source, Multilingual Information Extraction and Summarization. Theory and Applications of Natural Language Processing, pp. 23–49. Springer (2012)
9. Pivovarova, L., Du, M., Yangarber, R.: Adapting the PULS event extraction framework to analyze Russian text. In: At ACL: 4th Biennial Workshop on Balto-Slavic Natural Language Processing, Sofia, Bulgaria, pp. 100–109 (2013)
10. Steinberger, R.: A survey of methods to ease the development of highly multilingual text mining applications. Lang. Resources & Evaluation 46, 155–176 (2012)

Nordic Music Genre Classification Using Song Lyrics

Adriano A. de Lima, Rodrigo M. Nunes,
Rafael P. Ribeiro, and Carlos N. Silla Jr.

Computer Music Technology Laboratory
Federal University of Technology of Parana (UTFPR)
Av. Alberto Carazzai, 1.640 - 86300-000
Cornélio Procópio, PR, Brazil
{adrianolima.aal,kyllopardiun,rpr.rafa,carlos.sillajr}@gmail.com
http://music.cp.utfpr.edu.br

Abstract. Lyrics-based music genre classification is still understudied within the music information retrieval community. The existing approaches, reported in the literature, only deals with lyrics in the English language. Thus, it is necessary to evaluate if the standard text classification techniques are suitable for lyrics in languages other than English. More precisely, in this work we are interested in analyzing which approach gives better results: a language-dependent approach using stemming and stopwords removal or a language-independent approach using n-grams. To perform the experiments we have created the Nordic music genre lyrics database. The analysis of the experimental results shows that using a language-independent approach with the n-gram representation is better than using a language-dependent approach with stemming. Additional experiments using stylistic features were also performed. The analysis of these additional experiments has shown that using stylistic features combined with the other approaches improve the classification results.

Keywords: Lyrics Classification, Multi-language text classification, Music Genre Classification.

1 Introduction

Multimedia technologies have been an important field in many research areas, for example, Music Information Retrieval (MIR) has focused in important research directions such as music similarity retrieval, musical genre classification or music analysis and knowledge representation [1]. The content of the multimedia data might be audio, album covers, social and cultural data, music videos or song lyrics. In [2–6] it has been shown that song lyrics are relevant to assist or substitute other approaches for the task of music genre classification using standard text classification techniques. However, according to [7] the use of lyrics generally depends on language processing tools, that might not be available for songs whose lyrics are not in English (e.g. for Nordic music).

E. Métais, M. Roche, and M. Teisseire (Eds.): NLDB 2014, LNCS 8455, pp. 89–100, 2014.

The main objective of this paper is to verify which approach is more suitable for the lyrics-based music genre classification when dealing with lyrics that are not in the English language. For this purpose we have employed two standard approaches for lyrics processing (presented in Section 2): the first approach uses language-dependent resources while the second approach is language-independent. In order to evaluate both approaches on non-English Lyrics, we developed the Nordic Music Genre Lyrics Database (presented in Section 3). Our computation experiments with the two different approaches for the task of lyrics-based music genre classification and their results are shown in Section 4. In Section 5 we present the conclusions and future research directions of this work.

2 Lyrics Processing

An approach commonly employed for text processing is bag-of-words wherein each document is represented as word vectors that occurs into document, this approach is necessary to standardize the documents into a fixed format understood by the classifier. Thus, in this section we present the text processing techniques used to build the bag-of-words model applied in this work for the task of lyrics-based music genre classification. In Section 2.1 we describe the lyrics pre-processing steps. In Section 2.2 we describe the optional step of stopwords removal. Note that although stopwords are language-dependent, we have allowed this optional step in both approaches for the sake of completeness in our experiments. In Section 2.3 we describe the n-gram approach, which is a language-independent approach, for representing documents as word vectors. In Section 2.4 we describe the language-dependent stemming approach for representing documents as word vectors. In Section 2.5 we provide a brief explanation of the Term Frequency-Inverse Document Frequency (TF-IDF) technique and how it is used with the approaches presented in sections 2.3 and 2.4. In Section 2.6 we present an alternative approach for extracting features from lyrics, known as stylistic features.

2.1 Pre-processing

This step removes accents, punctuation, special characters, breaking words with hyphens, treatment of capital letters and equivalence of characters process. This procedure is necessary because we are dealing with non-usual lyrics, special attention is needed to some special characters present in the Nordic lyrics that should be mapped to specific characters in order to maintain their meaning. Table 1 presents the mapping between the Nordic languages special characters and their equivalent form.

2.2 Stopwords Removal

The stopwords are very common words (i.e. articles and prepositions) that carry little semantic meaning to the text, these irrelevant words are included in a

Table 1. Special Characters Mapping for the Nordic Languages

Language(s)	Special Character	Equivalent Character
Danish, Norwegian	æ	ae
Danish, Norwegian	ø	oe
Danish, Norwegian, Swedish	å	aa
Swedish	ö	oe
Swedish	ä	ae

pre-defined list, called stopwords list or stoplist[1] [8]. In this work we have used one stoplist for each Nordic language with 114 stopwords for Swedish, 176 for Norwegian and 94 for Danish.

In [9, 10] the authors studied the impact of stopwords for text and music mood classification. However, as we are interested in understanding the impact of these stopwords in the music genre classification using Nordic lyrics, we have analyzed whether it is advantageous or not to remove them before performing the task of lyrics-based music genre classification.

2.3 N-Grams

This technique transforms words into n-grams. A n-gram is a sequence of n characters (adding the special character "_" to denote the beginning and ending of the word), for example, given the term "text": for $n=2$, the representative attributes would be "_t", "te", "ex", "xt" and "t_"; for $n=3$, "_te", "tex", "ext" and "xt_"; and for $n=4$, "_tex", "text" and "ext_" [11].

In distinct language cases this technique is generally used when languages have a common root. Thus, the variation of the some words does not influence the morphemes that are common between languages. Therefore, this approach allows to infer the morphemes dividing the word into a character sequence of fixed size, n, which is predefined according with application context.

2.4 Stemming

The stemming approach is used to obtain the root form of all the words from a set of lyrics. For example, given the words "computing" and "computer", the result of a stemming algorithm could be "comput" [12]. In [13] it has been shown that this technique is largely used in the text classification domain with the objective of reducing the dimensionality. It is also a very fast procedure. It is important to highlight that the stemming algorithms are language specific and therefore may not be available to some languages [14].

2.5 TF-IDF

There are many methods to calculate the weight of each term (e.g., Weighted Inverse Document Frequency (WIDF) [15] and Term Relevance [16]), being the

---

[1] The stoplists were obtained from http://snowball.tartarus.org/

TF-IDF one of most commonly employed [17]. The advantage of TF-IDF, is the trade-off between how often a term occurs (Term frequency) and in how many documents the term occurs. The rationale being that a term that occurs frequently in a given document may be important but only if it does not occur in every other document as well. The formula for computing the TF-IDF is presented in Equation (1).

$$TFIDF = TF \times \log \frac{nD}{nDF} \tag{1}$$

where:

- TF is frequency of the term in a document, in other words, the number times that term occurs in a document;
- nD is the number of documents used;
- nDF is the number of documents where occurs the representation of that term;

2.6 Stylistic Features

The use of stylistic features for lyrics-based music genre classification was originally proposed in the work of [2]. In this work they have proposed simple counts of special punctuations (e.g., "!", "-", "?"), occurrences of digits, words and unique words per line, unique words ratio and characters per word. The main motivation behind using stylistic features is to capture the "interjection words" in the lyrics,

Table 2. Subset of the stylistic features proposed in [10] used in this work

Feature	Definition
interjection words	normalized frequency of "hey", "ooh", "yo", "uh", "ah", "yeah", "hej", "vad", "hejsan", "oj", "aj", "ja", "hei"
special punctuation marks	normalized frequency of "!", ".", "?", "-", ","
NUMBER	normalized frequency of all non-year numbers
No. of words	total number of words
No. of uniqWords	total number of unique words
repeatWordRatio	(No. of words - No. of uniqWords) / No. of words
avgWordLength	average number of characters per word
No. of lines	total number of lines
No. of uniqLines	total number of unique lines
No. of blankLines	number of blank lines
blankLineRatio	No. of blankLines / No. of lines
avgLineLength	No. of words / No. of lines
stdLineLength	standard deviation of number of words per line
uniqWordsPerLine	No. of uniqWords / No. of lines
repeatLineRatio	(No. of lines - No. of uniqLines) / No. of lines
avgRepeatWordRatioPerLine	average repeat word ratio per line
stdRepeatWordRatioPerLine	standard deviation of repeat word ratio per line

special punctuations and to compute simple text statistics. In this work we employ a subset of the stylistic features proposed for the task of lyrics-based music mood classification in [10]. We have used this subset because it complements the original stylistic features with new measures and also due to the improved results obtained in [2]. Note that in this work we have not used the stylistic features related to the song recordings (e.g. the number of words per minute or the number of lines per minute). Table 2 presents the stylistic features used in this work.

The main advantage of the stylistic features is that they use simple statistical measures based on word or character frequencies. In the context of this work, this technique was used in order to evaluate its suitability for the task of lyrics-based music genre classification.

3   Experimental Settings

In this section we present the process to create lyrics database with its characteristics (Sections 3.1 and 3.2) and the classifier used in the experiments (Section 3.3).

3.1   Nordic Lyrics Database Creation

The Nordic Music Genre Classification Lyrics Database was created using the following procedure: In the first stage we applied a questionnaire to some residents of Denmark, Norway and Sweden. In this questionnaire we asked the participants to inform their favorite songs along with the songs title, performing artist, music genre and language. Note that we imposed a restriction on the questionnaire that only songs sung in either Danish, Swedish or Norwegian were accepted. In the second stage, we searched the web manually to acquire the lyrics for our experiments. The main motivation behind the development of this novel database is to extend the analysis of lyric-based music genre classification for languages other than English.

The final version of the Nordic Music Genre Classification Lyrics Database contains 1,513 song lyrics divided in five genres (Dance: 265, Pop Rock: 263, Rap: 100, Pop: 357 and Rock: 528) distributed in three Nordic languages (Fig. 1 shows the distribution of song lyrics according to each language while Figures 2 to 6 show the word clouds for each genre using the 150 words with the highest values of TF-IDF and a list of the top 10 terms translated into the English language.). The music genres were labeled following the definition shown in [18]: "a kind of music, as it is acknowledged by a community for any reason or purpose or criteria". It is noteworthy that in the Nordic lyrics they contain some terms from the English language.

3.2   Dataset Details

The quantity of attributes generated by representation for Nordic Music Genre Classification Lyrics Database using stopwords (represented by Sw) or after of stopwords removal are shown in Table 3.

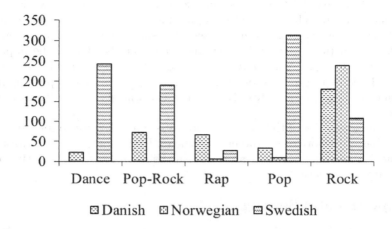

Fig. 1. Distribution of song lyrics by genres according to each Nordic language

Term	TF-IDF	Translation
jag	0.153	i (swedish)
aer	0.150	to be (verb)
i	0.137	in/of/at (preposition)
du	0.124	you (personal pronoun)
foer	0.121	before
och	0.120	and (swedish)
ska	0.120	shall
igen	0.107	again
vill	0.107	will
naer	0.105	near or when/that/as

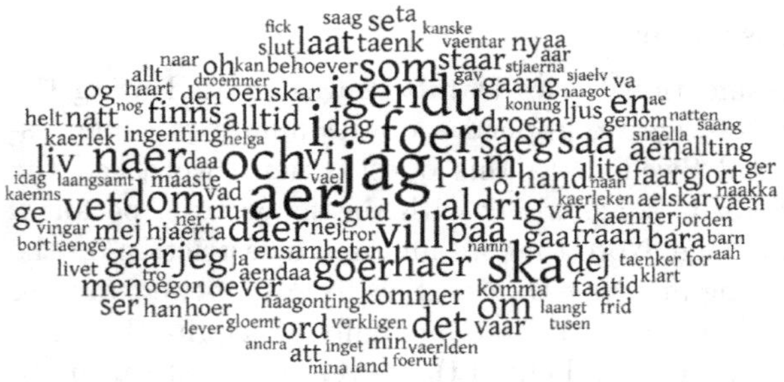

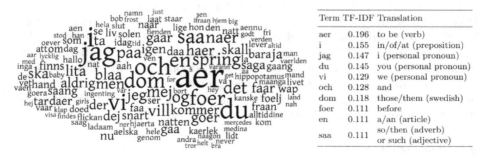

Fig. 2. Word cloud with the 150 words with the highest TF-IDF and a list of the top 10 terms translated into the English language for the Pop music genre

Term	TF-IDF	Translation
aer	0.196	to be (verb)
i	0.155	in/of/at (preposition)
jag	0.147	i (personal pronoun)
du	0.145	you (personal pronoun)
vi	0.129	we (personal pronoun)
och	0.128	and
dom	0.118	those/them (swedish)
foer	0.111	before
en	0.111	a/an (article)
saa	0.111	so/then (adverb) or such (adjective)

Fig. 3. Word cloud with the 150 words with the highest TF-IDF and a list of the top 10 terms translated into the English language for the Pop Rock music genre

Term	TF-IDF	Translation
aer	0.181	to be (verb)
jag	0.151	i (swedish)
saa	0.149	so/then or such
i	0.148	in/of/at (preposition)
eg	0.140	i (new norwegian)
e	0.138	to be (dialect form)
va	0.130	to be (informal)
det	0.117	the (article) or it/that (prenom)
jeg	0.117	i (norwedian/danish)
te	0.107	to (norwedian dialect)

Fig. 4. Word cloud with the 150 words with the highest TF-IDF and a list of the top 10 terms translated into the English language for the Rock music genre

Term	TF-IDF	Translation
i	0.181	in/of/at
jeg	0.163	i (personal pronoun)
bellstar	0.163	belle star (soap opera star)
baby	0.130	baby ("pretty" girl)
ah	0.126	onomatopoeia
saa	0.126	so/then or such
aer	0.126	to be (verb)
paa	0.117	on/at/in/upon/ into (preposition)
og	0.109	and
naar	0.097	when

Fig. 5. Word cloud with the 150 words with the highest TF-IDF and a list of the top 10 terms translated into the English language for the Rap music genre

Term	TF-IDF	Translation
aer	0.162	to be (verb)
jag	0.141	i (swedish)
du	0.135	you (personal pronoun)
i	0.133	in/of/at (preposition)
och	0.125	and (swedish)
ich	0.118	i (german)
vill	0.115	will
foer	0.107	before
dej	0.106	you (as an object in the setence)
blue	0.102	blue (english)

Fig. 6. Word cloud with the 150 words with the highest TF-IDF and a list of the top 10 terms translated into the English language for the Dance music genre

Table 3. Number of attributes by representation

Representation	Number of Attributes	
	With Sw	Without Sw
Stemming	20,962	20,704
2-Grams	1,087	1,079
3-Grams	9,945	9,829
4-Grams	36,741	35,705

An analysis of the different representations presented in Table 3 shows that the number of attributes grows considerably as we increase the parameter n for the n-grams representation. The stemming representation contains more than 20,000 features which is a high dimensionality when compared to the bigram and trigram representations. Furthermore, according to the details presented in Table 3, the optional step of stopwords removal has little impact in filtering out the common words in Nordic Lyrics. This might be due to the fact that music composers aims at creating new song lyrics by using seldom used words. The stylistic features on other hand produce a vector of low dimensionality with 17 only numeric values that can be used alone or in conjunction with other attributes.

3.3   Classifier

In this work we have employed the Support Vector Machine (SVM) classifier for the task of lyrics-based music genre classification. The SVM classifier is a supervised machine learning algorithms that uses hyperplanes to separate two classes by mapping the training examples [19]. With the objective of classify multiple classes we use the Sequential Minimal Optimization (SMO) algorithm[2] for training SVMs with default parameters and normalized features [21].

In [22] it has been shown that SVMs present three main advantages: they do not require complex tuning of parameters, they can generalize small training set and they are suited to learning in high dimensional spaces. Furthermore, the SVM classifier is gaining attention in MIR mainly for its performance and better results compared with other standard classifiers such as Decision Trees (DT), K-Nearest Neighbors (KNN) and Random Forests (RF) [23]. Among other papers that used SVM and its derivations for music genre classification task are [2–6].

4   Computational Results

In this section we are interested in answering the following questions by using controlled experiments: When dealing with Nordic lyrics which document representation should be used (Section 4.1)? Is a language-dependent representation

---

[2] Implementation from the Weka Data Mining Tool [20].

(using language tools such as stemmers) better than language-independent representation (using n-grams)? What is the impact of stopwords in music genre classification (Section 4.2)? Is it possible to combine the stylistic features with the other representations to obtain better results (Section 4.3)? For these purposes we perform our experiments using the Nordic lyrics music genre classification dataset presented in Section 3.1. To evaluate the results in this work we have used the F_1-score (which is the harmonic mean of precision and recall measures) with the ten-fold cross-validation procedure.

4.1   Which Representation to Use?

The results for the individual representations without removing the stopwords are presented in Table 4. Analyzing the results we had three main findings: First, the n-gram representation is better than stem in all parametrization. This is an interesting result as the n-gram representation can be used with any language without needing any language-specific resources. Second, by increasing the value of n in the n-gram representation it is possible to improve the classification results. Third, the use of stylistic features provides an interesting result as it achieves similar classification results with the stemming approach but using only 17 features.

Table 4. Results for the individual representations without stopwords removal

Music Genre	Stem	2-Grams	3-Grams	4-Grams	Stylistic
Dance	13.4	42.0	66.1	75.7	43.4
Pop Rock	14.8	62.5	67.1	100	00.7
Rap	47.2	44.0	97.5	47.4	48.6
Pop	37.4	63.0	77.0	80.1	21.1
Rock	62.5	74.7	84.5	98.8	56.3
Overall	38.7	62.9	77.4	87.1	35.6

4.2   What Is the Impact of Stopwords in Lyrics Classification?

When dealing with lyrics is still not clear if the stopwords should be removed. For this reason, we present the results for the individual representations with stopwords removal in Table 5. The analysis of the results presented in Table 5 shows that removing the stopwords from the lyrics harms the classification results for almost every representation.

4.3   Combining the Stylistic Features with the Other Representations

In this subsection we are interested in verifying if it is possible to improve the classification results by combining the stylistic features with the other representations. The main motivation for this combination is that the stylistic features

Table 5. Results for the individual representations with stopwords removal

Music Genre	Stem	2-Grams	3-Grams	4-Grams	Stylistic
Dance	10.5	42.1	68.4	75.3	43.3
Pop Rock	14.9	59.8	67.5	100	00.7
Rap	40.3	39.8	97.5	50.6	50.5
Pop	31.4	64.7	78.3	80.7	09.9
Rock	59.2	73.8	84.7	99.0	55.0
Overall	35.1	62.6	78.4	87.5	32.6

obtained individual results that were on par with the results obtained by using the stemming representation. Furthermore, the stylistic features are a language independent approach that can be easily computed and has only 17 attributes. The results of this experiment is presented in Table 6.

The analysis of the results in Table 6 shows that by combining the stylistic features with the stemming representation the classification results was improved from 38.7% (using only stemming) to 47%. This result is still worse than using any of the n-grams representation alone. Furthermore, when combining the n-grams representation with the stylistic features there is a small gain in classification results. More precisely from 62.9% to 64.2% for bigrams; from 77.4% to 78.1% for trigrams and from 87.1% to 87.9% for quadgrams.

Table 6. Results for the combination of the stylistic features with the other representations and without stopwords removal

Music Genre	STY+Stem	STY+2-Grams	STY+3-Grams	STY+4-Grams
Dance	38.6	51.1	68.8	74.5
Pop Rock	22.2	60.3	66.9	100
Rap	50.9	47.3	97.5	57.0
Pop	43.6	65.2	78.3	81.3
Rock	65.0	75.4	84.6	99.1
Overall	47.0	64.2	78.1	87.9

5 Conclusions

In this paper we have investigated which text classification approach is more suitable for the task of non-English lyrics-based music genre classification. In our experiments we have employed the stemming and n-grams based approaches. The stemming approach uses language-dependent resources which may not be available for some languages while using n-grams provides a language-independent approach. In addition to these approaches we have also studied the impact of using stylistic features.

In order to evaluate these approaches on non-English lyrics we developed a novel dataset. This dataset is the Nordic lyrics music genre classification dataset and it contains 1,513 lyrics from three different Nordic languages (Swedish, Norwegian and Danish) and from five different music genres (Dance, Pop Rock, Rap, Pop and Rock).

Our experimental results with the Nordic lyrics music genre classification dataset suggests that the use of the n-grams approach is more suitable for the task than the use of the stemming approach or the stylistic features alone. This is an interesting result as the n-grams approach is a language independent approach. Furthermore, we have performed experiments with bigrams, trigrams and quadgrams and the best results were obtained by using the quadgrams representation.

Additional experiments were also performed in order to evaluate the impact of stopwords removal in the task of lyrics classification. This is important for two reasons: First, it is not clear if stopwords should be removed when dealing with lyrics; Second, stopwords are a language specific resource and they may not be available for some languages. Based on our experimental results our suggestion is that the step of stopword removal is not performed when dealing with lyrics-based music genre classification.

The last contribution of the paper was to verify whether it was beneficial or not to combine the stylistics features with the other approaches used in this work. The analysis of our experimental results suggests that they should be used with other representations as the stylistic features always improve the classification results.

As a future research directions we plan on creating other non-English lyrics datasets in order to perform extensive experiments in the task of non-English lyrics-based music genre classification and also to perform additional experiments using other text classification features, such as the ones used in [24, 25].

References

1. Orio, N.: Music retrieval: a tutorial and review. Foundations and Trends in Information Retrieval 1(1), 1–90 (2006)
2. Mayer, R., Neumayer, R., Rauber, A.: Combination of audio and lyrics features for genre classification in digital audio collections. In: Proceedings of the 16th ACM International Conference on Multimedia, pp. 159–168 (2008)
3. Mayer, R., Neumayer, R., Rauber, A.: Rhyme and style features for musical genre classification by song lyrics. In: Proceedings of the 9th International Conference on Music Information Retrieval, pp. 337–342 (2008)
4. Mayer, R., Neumayer, R.: Multi-modal analysis of music: A large-scale evaluation. In: Proceedings of the Workshop on Exploring Musical Information Spaces, pp. 30–35 (2009)
5. Mayer, R., Rauber, A.: Building ensembles of audio and lyrics features to improve musical genre classification. In: Proceedings of the International Conference on Distributed Framework and Applications, pp. 1–6 (2010)
6. Mayer, R., Rauber, A.: Musical genre classification by ensembles of audio and lyrics features. In: Proceedings of International Conference on Music Information Retrieval, pp. 675–680 (2011)

7. Silla Jr., C.N., Koerich, A.L., Kaestner, C.A.A.: Improving automatic music genre classification with hybrid content-based feature vectors. In: Proceedings of the 2010 ACM Symposium on Applied Computing, pp. 1702–1707 (2010)
8. El-Khair, I.A.: Effects of stop words elimination for arabic information retrieval: a comparative study. International Journal of Computing & Information Sciences 4(3), 119–133 (2006)
9. Yu, B.: An evaluation of text classification methods for literary study. Literary and Linguistic Computing 23(3), 327–343 (2008)
10. Hu, X., Downie, J.S.: Improving mood classification in music digital libraries by combining lyrics and audio. In: Proceedings of the 10th Annual Joint Conference on Digital Libraries, pp. 159–168 (2010)
11. Cavnar, W.B., Trenkle, J.M.: N-gram-based text categorization. In: Proceedings of the 3rd Annual Symposium on Document Analysis and Information Retrieval, pp. 161–175 (1994)
12. Porter, M.F.: An algorithm for suffix stripping. Program: Electronic Library and Information Systems 14, 130–137 (1980)
13. Sebastiani, F.: Machine learning in automated text categorization. ACM Computing Surveys 34(1), 1–47 (2002)
14. Porter, M.F.: Snowball: A language for stemming algorithms, http://snowball.tartarus.org/texts/introduction.html
15. Tokunaga, T., Makoto, I.: Text categorization based on weighted inverse document frequency. Technical report, Tokyo Institute of Technology (1994)
16. Wu, H., Salton, G.: A comparison of search term weighting: Term relevance vs. inverse document frequency. In: Proceedings of the 4th Special Interest Group on Information Retrieval, pp. 30–39 (1981)
17. Salton, G., Buckley, C.: Term-weighting approaches in automatic text retrieval. Information Processing & Management 24(5), 513–523 (1988)
18. Fabbri, F.: Browsing music spaces: Categories and the musical mind (1999)
19. Burges, C.J.C.: A tutorial on support vector machines for pattern recognition. Data Mining and Knowledge Discovery 2(2), 121–167 (1998)
20. Witten, I.H., Frank, E.: Data Mining: Practical machine learning tools and techniques. Morgan Kaufmann (2005)
21. Platt, J.C.: Fast training of support vector machines using sequential minimal optimization. In: Advances in Kernel Methods, pp. 185–208 (1999)
22. Dhanaraj, R., Logan, B.: Automatic prediction of hit songs. In: Proceedings of International Conference on Music Information Retrieval, pp. 488–491 (2005)
23. Laurier, C., Grivolla, J., Herrera, P.: Multimodal music mood classification using audio and lyrics. In: Proceedings of the 7th International Conference on Machine Learning and Applications, pp. 688–693 (2008)
24. Koppel, M., Schler, J., Argamon, S.: Computational methods in authorship attribution. Journal of the American Society for Information Science and Technology 60(1), 9–26 (2009)
25. HaCohen-Kerner, Y., Beck, H., Yehudai, E., Rosenstein, M., Mughaz, D.: Cuisine: Classification using stylistic feature sets and/or name-based feature sets. Journal of the American Society for Information Science and Technology 61, 1644–1657 (2010)

Infographics Retrieval: A New Methodology

Zhuo Li[1], Sandra Carberry[1], Hui Fang[2], Kathleen F. McCoy[1],
and Kelly Peterson[1]

[1] Department of Computer and Information Science,
[2] Department of Electrical and Computer Engineering,
University of Delaware

Abstract. Information graphics, such as bar charts and line graphs, are a rich knowledge source that should be accessible to users. However, techniques that have been effective for document or image retrieval are inadequate for the retrieval of such graphics. We present and evaluate a new methodology that hypothesizes information needs from user queries and retrieves infographics based on how well the inherent structure and intended message of the graphics satisfy the query information needs.

Keywords: Graph retrieval, natural language query processing.

1 Introduction

Information graphics (infographics), such as the one in Figure 1, are effective visual representations of complex information. Moreover, the overwhelming majority of information graphics from popular media appear to be designed to convey an intended message. For example, the intended message of the graphic in Figure 1 is ostensibly that Toyota has the highest profit among the car manufacturers listed. Although much research has addressed the retrieval of documents, very little attention has been given to the retrieval of infographics. But infographics are an important knowledge resource that should be accessible from a digital library.

Suppose that one is writing a government report about the success of Japanese car companies and poses the following query:

Q_1: *"How does the net profit of Toyota compare to other car manufacturers?"*

A graphic satisfying this information need would presumably depict *net profit* on the dependent axis and *Toyota* along with other *car manufacturers* on the independent axis; in addition, the graphic would compare Toyota's net profit against those of the other car manufacturers. The infographic in Figure 1, which appeared in zmetro.com, is such a graphic. When Q_1 was entered into major commercial image search engines, none of the highest ranked graphics were comparing car manufacturers according to their net profit, and thus they would not satisfy the query's information need.

Techniques that have been effective for document or image retrieval are inadequate for the retrieval of information graphics. Current search engines employ

E. Métais, M. Roche, and M. Teisseire (Eds.): NLDB 2014, LNCS 8455, pp. 101–113, 2014.

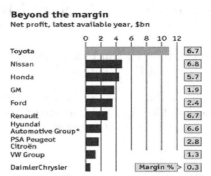

Fig. 1. An Example Infographic

strategies similar to those used in document retrieval, relying primarily on the text surrounding a graphic and web link structures. But the text in the surrounding document generally does not refer explicitly to the infographic or even describe its content [5]. An obvious extension would be to collect all the words in an infographic and use it as a bag of words. However, the text in graphics is typically sparse and we will show that this approach is insufficient. Content-based image retrieval (CBIR) has focused on extracting visual features or retrieving images that are similar to a user specified query image. Although CBIR can find bar charts and line graphs, a user with an information need is not just, if at all, seeking visually similar graphics. Moreover, since images are free-form with relatively little inherent structure, it is extremely difficult to determine what is conveyed by an image, other than to list the image's constituent pieces.

Infographics, on the other hand, have structure: the independent axis depicts a set of entities (perhaps ordinal entities in the case of a line graph) and the dependent axis measures some criteria for each entity. In addition, the graphic designer constructs the graphic using well-known communicative signals (such as coloring an entity in the graphic differently from other entities) to convey an intended message. Current retrieval mechanisms, both those used for document retrieval and those used for image retrieval, have ignored a graphic's structure and message, and thus have been ineffective in retrieving infographics.

We propose a novel methodology for retrieving infographics in response to a user query. Our approach analyzes the user query to hypothesize the desired content of the independent and dependent axes of relevant infographics and the high-level message that a relevant infographic should convey. It then ranks candidate graphics using a mixture model that takes into account the textual content of the graphic, the relevance of its axes to the structural content requested in the user query, and the relevance of the graphic's intended message to the information need (such as a comparison) identified from the user's query. We currently focus on static simple bar charts and line graphs; in the future, our methodology will be extended to more complex infographics.

This paper presents our new methodology for retrieving infographics, together with experiments which validate our approach. Section 2 outlines the problem in more detail. Section 3 presents our methodology and describes how relevance of each graphic component to a user query is measured. Section 4 presents experimental results showing the significant improvement our methodology makes over a baseline approach that uses simple text retrieval techniques.

2   Problem Formulation

Our research is currently limited to two kinds of infographics, simple bar charts and single line graphs that convey a two dimensional relationship between the independent axis and the dependent axis, which we will refer to respectively as x-axis and y-axis. Let an infographic be G. Given a digital library D of such infographics, we will present a ranking methodology for all $G \in D$ according to the relevance of each graphic G to a query Q, denoted $R(Q, G)$. We assume that in our digital library, each infographic G is stored with its original image together with an XML representation specifying the structural components of the graphic as described in Section 2.1 and the message components of the graphic as described in Section 2.2. This paper is not concerned with the computer vision problem of parsing a graph to recognize its bars, labels, colors, text, etc.; other efforts, such as the work in [6,13], are addressing the processing of electronic images such as bar charts and line graphs.

2.1   Structural Content of Infographics

Our approach takes into account three structural components of each infographic G: 1) text that appears in a graphic's caption or sub-caption, denoted G_c; 2) the content of the independent axis (referred to as *x-axis*), denoted as G_x, consisting of the labels on the x-axis, such as the names of the car manufacturers in Figure 1; 3)the content of the dependent axis (referred to as *y-axis*), denoted as G_y.

Determining G_y is not straightforward. Infographics often do not explicitly label the y-axis with what is being measured. For example, *net profit* is being measured on the y-axis in Figure 1 but the y-axis itself is unlabeled. Previous work on our project developed a system that utilized a set of heuristics to extract information from other graph components and meld them together to form a y-axis descriptor [9]. We assume that the XML representation of each infographic in our digital library contains all three structural components, $G_{Struct} = \{G_x, G_y, G_c\}$.

2.2   Message Content of Infographics

Infographics in popular media generally have a high-level message G_{IM} that they are intended to convey. For example, Figure 1 conveys a message in the *Max* category, namely *Toyota* has the highest net profit among the listed car manufacturers. Previous work on our project [10,27] identified a set of 17 categories of intended messages that could be conveyed by simple bar charts and line graphs. For example, a *Rank* message conveys the rank of an entity with respect to some criteria (such as profit) whereas a *Relative-Difference* message compares two specific entities. The entity being ranked or the entities being compared are referred to as *focused entities* G_{fx} since the graphic is focused on them.

Both studies [10,27] identified communicative signals that appear in graphics and help to convey the graphic's intended message. For example, salience of an entity in a graphic might be conveyed by coloring the bar differently from other bars (such as the bar for Toyota in Figure 1), thereby suggesting that it plays a significant role in the graphic's high-level message. These communicative signals were entered as evidence in a Bayesian network that hypothesized the graphic's intended message, both the category of intended message (such as *Rank*) and any focused entities that serve as parameters of the message.

We assume that each infographic G in our digital library is stored along with its message components, $G_{msg} = \{G_{IM}, G_{fx}\}$, its structural components $G_{struct} = \{G_x, G_y, G_c\}$, and a bag of words G_t of all the words in graphic G.

3   Retrieval Methodology

Given a query, our retrieval methodology first analyzes the query to identify the requisite characteristics of infographics that will best satisfy the user's information need. Then the infographics in our digital library are rank-ordered according to how well they satisfy this information need as hypothesized from the user's query. Section 3.1 discusses the analysis of user queries, Section 3.2 discusses the fast preselection of a set of candidate infographics and Section 3.3 discusses the rank-ordering of infographics in response to a user query.

3.1   Natural Language Query Processing

Our vision is that since the graphics have structure, the users whose particular information needs could be satisfied by infographics will formulate their queries to indicate the requisite structure of the desired graphics. Thus our methodology uses full-sentence user queries so that the semantics of the query can be analyzed to identify characteristics of relevant graphics. Consider the following two queries which contain similar keywords but represent different information needs:

Q_2: *Which Asian countries have the most endangered animals?*
Q_3: *Which endangered animals are found in the most Asian countries?*

Both queries request graphics that convey the entities that have the largest value with respect to what is measured on the y-axis. However, query Q_2 is asking for an infographic comparing *Asian countries* on the x-axis according to their number of endangered animals on the y-axis, while query Q_3 is asking for a comparison of different endangered animals on the x-axis according to the number of Asian countries in which they reside on the y-axis. Our methodology utilizes our previous work that extracted clues from a query Q and used these clues as attributes in a learned model that hypothesized from Q the requisite content Q_x of the x-axis, the requisite content Q_y of the y-axis, and the noun

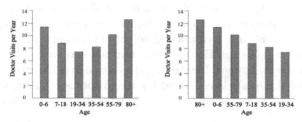

Fig. 2. Graphics Displaying the Same Data with Different Messages

words that do not belong on any of these axes Q_n [19,20]. Our retrieval methodology treats these axes contents separately instead of treating a query as a bag of words, thereby recognizing the difference between queries such as Q_2 and Q_3.

Similarly, consider the query:

Q_4:*How does the number of doctor visits per year change with a person's age?*

Although both graphics in Figure 2 contain the same data, that is, *a person's age* on the x-axis and *the number of doctor visits* on the y-axis, the changing trend in doctor visits (as requested by the query) is more discernible from the graphic on the left than from the graphic on the right which ranks age ranges in terms of number of doctor visits. This correlates with Larkin and Simon's observation [18] that graphics may be informationally equivalent (that is, they contain the same information) but not computationally equivalent (that is, it may be more difficult to perceive the information from one graphic than from the other). The model developed in our previous work [19] also processed a user query to identify the preferred category of intended message Q_{IM} and the focused entity Q_{fx} (if any) of relevant infographics.

In this paper, we process each user query Q using the learned models from our previous work [19,20] in order to 1) extract all of the query components that convey structural information, $Q_{struct} = \{Q_x, Q_y, Q_n\}$, and 2) identify the requisite message information that should be conveyed by the desired graphic $Q_{msg} = \{Q_{IM}, Q_{fx}\}$. In addition, we form a bag of words Q_t consisting of all words in query Q.

3.2   Infographics Preselection

As a first step for speeding-up infographic retrieval in a large digital library, we first preselect a subset of infographics that are loosely relevant to the words in a given user query. However, the words in a user query may differ from those appearing in a relevant graphic, especially since a graphic's text is typically sparse. Query expansion is a commonly used strategy in IR to improve retrieval performance [1,11,22] since it is effective in bridging the vocabulary gap between terms in a query and those in the documents. The basic idea is to expand the original query with semantically similar terms other than those explicitly given in the original query.

But infographic retrieval presents an additional problem. Consider a query such as *Which car manufacturer has the highest net profit?*. An infographic such as the one shown in Figure 1 displays a set of car manufacturers on the independent axis (Toyota, Nissan, Honda, etc..) but nowhere in the graphic does the term *car* or a synonym appear. Identifying the ontological categories, such as *car* or *automobile*, of these labels is crucial for infographics retrieval since the user query often generalizes the entities on the independent axis rather than listing them.

To tackle the sparsity problem, before storing a graphic in the digital library, we expand the text in the graphic using Wikimantic[4], a term expansion method that uses Wikipedia articles as topic concepts. Given a sequence s of the text in a graphic, Wikimantic extracts all Wikipedia articles and disambiguation articles whose titles contain a subsequence of s; each of these articles is viewed as a Wikimantic concept and is weighted by the likelihood that the concept generates sequence s. Wikimantic then builds a unigram distribution for words from the articles representing these weighted concepts. By expanding graph entities such as Toyota, Nissan, Honda, and GM through Wikimantic, words such as *car* or *automobile* are part of the produced unigram distribution — that is, as a side effect, the ontological category of the individual entities becomes part of the term expansion.

This expansion of the graphic components (as opposed to the typical expansion of the query) accomplishes two objectives: 1) it addresses the problem of sparse graphic text by adding semantically similar words, and 2) it addresses the problem of terms in the query capturing general classes (such as *car* or *automobile*) when the graphic instead contains an enumeration of members of that general class. If the expanded text of an infographic contains at least one of the query noun words, this infographic is preselected for further rank-ordering.

3.3   Methodology for Rank-Ordering Infographics

After preselecting a candidate pool P of infographics for a given query Q, our methodology measures the relevance $R(Q, G)$ of Q to each infographic $G \in P$. Our hypothesis is that graph retrieval should take into account the relevance of the structural and message components of an infographic to the requirements conveyed by the user's query. We consider the following relevance measurements, as depicted in Figure 3:

- X Axis Relevance $R(Q_x, G_x)$: relevance of the graphic's x-axis content G_x to the requisite x-axis content Q_x extracted from the user's query.
- Y Axis Relevance $R(Q_y, G_y)$: relevance of the graphic's y-axis content G_y to the requisite y-axis content Q_y extracted from the user's query.
- Intended Message Category Relevance $R(Q_{IM}, G_{IM})$: relevance of the category of intended message G_{IM} of the infographic to the category of intended message Q_{IM} preferred by the query.
- Intended Message Focused Entity Relevance $R(Q_{fx}, G_{fx})$ and $R(Q_{fx}, G_{nx})$: relevance of the graphic's focused entity G_{fx} (if any) to the focused entity

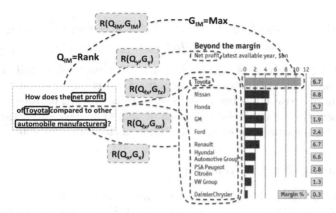

Fig. 3. Relevance Measurements for Component Approaches

Q_{fx} (if any) extracted from the user's query. In cases where Q_{fx} appears on the x-axis of a graphic but is not focused, such graphics may address the user's information need, though less so than if the graphic also focused on Q_{fx}. Therefore we also measure the relevance of the non-focused x-axis entities $G_{nx} \in G_x$ to the query focused entity Q_{fx} as $R(Q_{fx}, G_{nx})$.

We consider three mixture models which respectively capture structural relevance, message relevance, and both structural and message relevance. Since the results of query processing are not always correct, we add to each model a back-off relevance measurement $R(Q_t, G_t)$ which measures the relevance of all the words in the query to all the words in a candidate infographic. In addition, we include a baseline model that treats the words in the graphic and the words in the query as two bags of words and measures their relevance to one another.

Baseline-Model (bags of words): relevance of the bag of words in the query to the bag of words in a graphic, calculated by the following equation:

$$R_{baseline}(Q, G) = R(Q_t, G_t) \tag{1}$$

Model-1 (structural components): relevance of the structural components (the x-axis and the y-axis) computed by the following function:

$$R_1(Q, G) = \omega_0 \cdot R(Q_t, G_t) + \omega_1 \cdot R(Q_x, G_x) + \omega_2 \cdot R(Q_y, G_y) \tag{2}$$

Model-2 (message components): relevance of intended message components (message category and message focused entity, if any) computed by the following function:

$$R_2(Q, G) = \omega_0 \cdot R(Q_t, G_t) + \omega_3 \cdot R(Q_{IM}, G_{IM})$$
$$+\omega_4 \cdot R(Q_{fx}, G_{fx}) + \omega_5 \cdot R(Q_{fx}, G_{nx}) \tag{3}$$

Model-3 (both structural and message components): relevance of both structural and intended message components, computed by the following equation:

$$R_3(Q, G) = \omega_0 \cdot R(Q_t, G_t) + \omega_1 \cdot R(Q_x, G_x) + \omega_2 \cdot R(Q_y, G_y)$$
$$+\omega_3 \cdot R(Q_{IM}, G_{IM}) + \omega_4 \cdot R(Q_{fx}, G_{fx}) + \omega_5 \cdot R(Q_{fx}, G_{nx}) \tag{4}$$

The weighting parameters, ω_i, are learned using multi-start hill climbing to find a set of parameters that yields a local maximal retrieval evaluation metric. Such hill-climbing search has been used successfully to learn parameters in other problems where the available dataset is small [23]. The next subsections discuss how relevance is measured for each of the terms in the above relevance equations.

3.4   Measuring Textual Relevance

The relevance between the words from the query and words from the graphic, such as $R(G_t, Q_t)$, $R(G_x, Q_x)$, $R(G_y, Q_y)$, and $R(G_{fx}, Q_{fx})$, are textual relevances, measured by relevance function R_{text}. We use a modified version of Okapi-BM25 [12] for measuring textual relevance R_{text}:

$$R_{text}(Q_c, G_{c'}) = \sum_{w_i \in Q_c} log(\frac{|D| + 1}{gf_i + 1}) \cdot \frac{tf_i \cdot (1 + k_1)}{tf_i + k_1}$$

where Q_c is a query component and $G_{c'}$ is a graphic component, $|D|$ is the total size of our graphic collection, gf_i is the number of graphics that contain the word w_i, tf_i is the term frequency of w_i in $G_{c'}$, and k_1 is a parameter that is set to 1.2, a widely used value. This version of Okapi-BM25 has replaced the original inverse document frequency in Okapi with the regular inverse document frequency ($idf = log(\frac{|D|+1}{gf_i+1})$) to address the problem of negative idf. Our version of Okapi also does not take graphic text length into consideration, since text in graphics usually have similar limited length; moreover, a graph component, such as the message focused entity or the dependent (y) axis, only consists of a noun entity and therefore normalizing the length of such a component does not have the same affect as for documents. Our version of Okapi also does not take query term frequency into consideration, since most terms in the query occur only once.

3.5   Intended Message Relevance Measurement

Intended message relevance measures the relevance of the category of the intended message (such as *Rank*) along with the message focused entity of a query to those of an infographic. For measuring the intended message category relevance R_{IM}, we condense the message categories in [10] and [27] into seven general categories of intended messages extracted from user queries:

– *Trend* messages: convey a trend over some ordinal entity. Note that while a graphic might convey a rising trend, a query would be much more likely merely to request the trend of some entity since the user would not know a priori whether the trend is rising, falling, or stable.

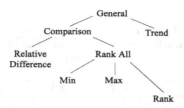

Fig. 4. Intended Message Category Similarity

- *Rank:* convey the rank of a specific entity with respect to other entities.
- *Min:* convey the entity that has the smallest value with respect to other entities.
- *Max:* convey the entity that has the largest value with respect to other entities.
- *Rank-all:* convey the relative rank of a set of entities
- *Rel-Diff:* convey the relative difference between two entities
- General: convey no specific message and just display data

We abstract a concept hierarchy containing the seven general intended message categories, as shown in Figure 4. Our methodology uses *relaxation* as the paradigm for ranking infographics according to how well an infographic's category of intended message G_{IM} satisfies the requisite intended message Q_{IM} extracted from the user query.

A six degree relevance measurement for R_{IM} is computed based on this hierarchy. When G_{IM} matches Q_{IM}, little perceptual effort is required for the user to get the message information he or she wants; this infographic is deemed fully relevant to the query in terms of message relevance. However, when G_{IM} differs from Q_{IM}, the amount of perceptual effort that the user must expend to satisfy his information need depends on G_{IM}. By moving up or down the intended message hierarchy from $Q_{IM} \rightarrow G_{IM}$, Q_{IM} is relaxed to match different G_{IM} with different degrees of penalties for the relaxation. The greater the amount of relaxation involved, the less message-relevant the infographic is to the query, and the more points penalized for message relevance.

At the top of the hierarchy is the *General* intended message category, which captures the least information message-wise. Message categories lower in the hierarchy contain more specific information. When Q_{IM} is lower in the hierarchy than G_{IM}, Q_{IM} requires more specific information than provided by G_{IM}. By relaxing $Q_{IM} \xrightarrow{up} G_{IM}$, perceptual effort is needed for the user to get the desired information; this infographic will be penalized for not having specific enough information. For example, consider two graphics, one whose intended message is the *Rank* of France with respect to other European countries in terms of cultural opportunities (and thus France is highlighted or salient in the graphic) and a second graphic whose intended message is just a ranking (category *Rank-all*) of all European countries in terms of cultural opportunities. If the user's query requests the rank of France with respect to other countries, then the first graphic matches the user's information need whereas the second graphic requires

Table 1. NDCG@10 Results

Query	Approach	Baseline	Model-1: Structural	Model-2: Message	Model-3: Structural and Message
Learned Model	exact match	0.3245	0.3766	0.3568	0.4168
	graph expansion	0.3905	0.4280	0.4191	0.4520
Hand Labeled	exact match	0.3245	0.4348	0.3881	0.4576
	graph expansion	0.3905	0.4782	0.4433	0.4866

a relaxation of message category from (Q_{IM} = Rank) $\xrightarrow{up}$ (G_{IM} = Rank-all); in this latter case, user effort is required to search for France among the countries listed and thus the second infographic is penalized for message relevance.

Limited space prevents further detail about the relaxation process.

4   Experimental Results and Discussion

4.1   Data Collection

A human subject experiment was performed to collect a set of 152 full sentence user queries from five domains. We used the collected queries to search on popular commercial image search engines to get more infographics from the same domain. This produced 257 infographics that are in the domain of the collected queries.

Each query-infographic pair was assigned a relevance score on a scale of 0-3 by an undergraduate researcher. A query-infographic pair was assigned 3 points if the infographic was considered highly relevant to the query and 0 points if it was irrelevant. Infographics that were somewhat relevant to the query were assigned 1 or 2 points, depending on the judged degree of relevance.

4.2   Experimental Results

In order to evaluate our methodology, we performed experiments in which we averaged together the results of 10 runs, with the Bootstrapping method [26] (a recommended evaluation method for small data sets) used to randomly select training and testing sets of queries. Normalized Discounted Cumulative Gain (NDCG) [17] is used to evaluate the retrieval result. It is between 0 and 1 and measures how well the rank-order of the graphs retrieved by our method agrees with the rank order of the graphs identified as relevant by our evaluator. We use a Student's t-test for computing significance.

The third column of Table 1 gives the NDCG@10 results for the baseline and the next three columns give the results for our three models (structural, message, and structural+message). The first two rows of Table 1 show results when the learned models from our previous work [19,20] are used to hypothesize requisite structural and message content from the user queries; since the learned models are not perfect, the last two rows of Table 1 show results when hand-labelled data is used.[1] Furthermore, the first and third rows show results when textual relevance is computed using exact match of query words with graph words, whereas the second and fourth rows give results when query words are matched with words in the expansion of the graph text via Wikimantic.

The experimental results, using the learned model for query analysis, show that utilizing structural relevance (Model-1) and message relevance (Model-2) each provide significantly better results than the baseline (p=0.001). Furthermore, the combination of structural and message relevance improves upon either alone (p=0.0005). The results also show that Wikimantic graph expansion improves retrieval performance significantly for all of the approaches (p≤0.005). Furthermore, the results using hand-labelled data are significantly better (p≤0.001) than the results using the learned models from our previous work to extract structural and message content from user queries, thereby indicating that improvement in these learned models should produce better retrieval results.

5   Related Work

Most infographic retrieval systems, such as SpringerImages and Zanran (http://www.springerimages.com, http://www.zanran.com), are based on graphics' textual annotations as in image retrieval [14], such as the graphics' names and surrounding text. There has been previous work on semi-structured data retrieval, such as xml data, that first measures the relevance of each semi-structured data element separately, and then combines the relevance measurements to form the overall measurement [15].

In this paper, we focus on natural language queries since they allow users to express their specific information need more clearly than keywords [19,24,3]. Previous work on verbose and natural language queries used probabilistic models and natural language processing techniques [2,21] to identify the key content in such queries. Research on retrieval of structured data, such as linked data and ontologies, also relies on the syntax and semantics of natural language queries [8]. However, attributes and relationships in the data are explicitly given in the ontologies, and the queries specify the desired attribute and/or relationship. In our work, the retrieved unit is a graphic as opposed to the value of an attribute, and extracting structural and message content from a query is more complex.

6   Conclusion and Future Work

This paper has presented a novel methodology for retrieving information graphics relevant to user queries; it takes into account a graphic's structural content and the graphic's high-level intended message. Our experimental results show that our methodology improves significantly over a baseline method that treats

---

[1] For the baseline method, the results are the same since structural and message information is not considered.

graphics and queries as single bags of words. To our knowledge, our work is the first to investigate the use of structural and message information in the retrieval of infographics. We are currently extending the variety of relevance features and exploring widely used learning-to-rank algorithms to achieve higher retrieval performance. We will also explore question-answering from relevant graphics.

Acknowledgements. This work was supported by the National Science Foundation under grant III-1016916 and IIS-1017026.

References

1. Arguello, J., Elsas, J.L., Callan, J., Carbonell, J.G.: Document representation and query expansion models for blog recommendation. In: Proc. ICWSM (2008)
2. Bendersky, M., Croft, W.B.: Discovering key concepts in verbose queries. In: Proc. of ACM SIGIR Conf. on Res. and Dev. in Information Retrieval, pp. 491–498 (2008)
3. Bendersky, M., Croft, W.B.: Analysis of long queries in a large scale search log. In: Proc. of the Workshop on Web Search Click Data, pp. 8–14 (2009)
4. Boston, C., Fang, H., Carberry, S., Wu, H., Liu, X.: Wikimantic: Toward effective disambiguation and expansion of queries. In: Data & Knowledge Engineering (2014)
5. Carberry, S., Elzer, S., Demir, S.: Information graphics: an untapped resource for digital libraries. In: Proc. of ACM SIGIR Conf. on Res. and Dev. in Information Retrieval, pp. 581–588 (2006)
6. Chester, D., Elzer, S.: Getting computers to see information graphics so users do not have to. In: Hacid, M.-S., Murray, N.V., Raś, Z.W., Tsumoto, S. (eds.) ISMIS 2005. LNCS (LNAI), vol. 3488, pp. 660–668. Springer, Heidelberg (2005)
7. Clark, S., Curran, J.: Wide-coverage efficient statistical parsing with ccg and log-linear models. Computational Linguistics 33(4), 493–552 (2007)
8. Damljanovic, D., Agatonovic, M., Cunningham, H.: Freya: An interactive way of querying linked data using natural language. In: García-Castro, R., Fensel, D., Antoniou, G. (eds.) ESWC 2011. LNCS, vol. 7117, pp. 125–138. Springer, Heidelberg (2012)
9. Demir, S., Carberry, S., Elzer, S.: Effectively realizing the inferred message of an information graphic. In: Proc. of the Int. Conf. on Recent Advances in Natural Language Processing, pp. 150–156 (2007)
10. Elzer, S., Carberry, S., Zukerman, I.: The automated understanding of simple bar charts. Artificial Intelligence 175(2), 526–555 (2011)
11. Escalante, H.J., Hernández, C., López, A., Marín, H., Montes, M., Morales, E., Sucar, L.E., Villaseñor, L.: Towards annotation-based query and document expansion for image retrieval. In: Peters, C., Jijkoun, V., Mandl, T., Müller, H., Oard, D.W., Peñas, A., Petras, V., Santos, D. (eds.) CLEF 2007. LNCS, vol. 5152, pp. 546–553. Springer, Heidelberg (2008)
12. Fang, H., Tao, T., Zhai, C.: A formal study of information retrieval heuristics. In: Proc. of ACM Conf. on Res. and Dev. in Information Retrieval, pp. 49–56 (2004)
13. Futrelle, R.P., Nikolakis, N.: Efficient analysis of complex diagrams using constraint-based parsing. In: Proc. of Int. Conf. on Document Analysis and Recognition, vol. 2, pp. 782–790 (1995)
14. Gao, Y., Wang, M., Luan, H., Shen, J., Yan, S., Tao, D.: Tag-based social image search with visual-text joint hypergraph learning. In: Proc. of Int. Conf. on Multimedia, pp. 1517–1520 (2011)

15. Hiemstra, D.: Statistical language models for intelligent xml retrieval. In: Blanken, H.M., Grabs, T., Schek, H.-J., Schenkel, R., Weikum, G. (eds.) Intelligent Search on XML Data. LNCS, vol. 2818, pp. 107–118. Springer, Heidelberg (2003)
16. Huang, W., Tan, C.L.: A system for understanding imaged infographics and its applications. In: Proc. of ACM Symposium on Document Engineering, pp. 9–18 (2007)
17. Järvelin, K., Kekäläinen, J.: Cumulated gain-based evaluation of ir techniques. ACM Trans. Inf. Syst. 20(4), 422–446 (2002)
18. Larkin, J.H., Simon, H.A.: Why a diagram is (sometimes) worth ten thousand words. Cognitive Science 11(1), 65–100 (1987)
19. Li, Z., Stagitis, M., Carberry, S., McCoy, K.F.: Towards retrieving relevant information graphics. In: Proc. of ACM SIGIR Conf. on Res. and Dev. in Information Retrieval, pp. 789–792 (2013)
20. Li, Z., Stagitis, M., McCoy, K., Carberry, S.: Towards finding relevant information graphics: Identifying the independent and dependent axis from user-written queries (2013), http://www.aaai.org/ocs/index.php/FLAIRS/FLAIRS13/paper/view/5939
21. Liu, J., Pasupat, P., Wang, Y., Cyphers, S., Glass, J.: Query understanding enhanced by hierarchical parsing structures. In: IEEE Workshop on Automatic Speech Recognition and Understanding, pp. 72–77 (2013)
22. Metzler, D., Cai, C.: Usc/isi at trec 2011: Microblog track. In: TREC (2011)
23. Metzler, D., Croft, W.B.: A markov random field model for term dependencies. In: Proc. of ACM Conf. on Res. and Dev. in Information Retrieval, pp. 472–479 (2005)
24. Phan, N., Bailey, P., Wilkinson, R.: Understanding the relationship of information need specificity to search query length. In: Proc. of ACM SIGIR Conf. on Res. and Dev. in Information Retrieval, pp. 709–710 (2007)
25. Savva, M., Kong, N., Chhajta, A., Fei-Fei, L., Agrawala, M., Heer, J.: Revision: Automated classification, analysis and redesign of chart images. In: Proc. of ACM Symposium on User Interface Software and Technology, pp. 393–402 (2011)
26. Tan, P., Steinbach, M., Kumar, V.: Introd. to Data Mining. Addison Wesley (2006)
27. Wu, P., Carberry, S., Elzer, S., Chester, D.: Recognizing the intended message of line graphs. In: Goel, A.K., Jamnik, M., Narayanan, N.H. (eds.) Diagrams 2010. LNCS (LNAI), vol. 6170, pp. 220–234. Springer, Heidelberg (2010)

A Joint Topic Viewpoint Model for Contention Analysis

Amine Trabelsi and Osmar R. Zaïane

Department of Computing Science, University of Alberta, Edmonton, Canada
{atrabels,zaiane}@ualberta.ca

Abstract. This work proposes an unsupervised Joint Topic Viewpoint model (JTV) with the objective to further improve the quality of opinion mining in contentious text. The conceived JTV is designed to learn the hidden features of arguing expressions. The learning task is geared towards the automatic detection and clustering of these expressions according to the latent topics they confer and the embedded viewpoints they voice. Experiments are conducted on three types of contentious documents: polls, online debates and editorials. Qualitative and quantitative evaluations of the models output confirm the ability of JTV in handling different types of contentious issues. Moreover, analysis of the preliminary experimental results shows the ability of the proposed model to automatically and accurately detect recurrent patterns of arguing expressions.

Keywords: Contention Analysis, Topic Models, Opinion Mining, Unsupervised Clustering.

1 Introduction

Our work fits into the lines of research that addresses the problem of enhancing the quality of opinion extraction from unstructured text found in social media platforms. Netizens use these novel media platforms to discuss and express their opinion over major socio-political events. These events, often, are the object of heated debates over a controversial or contentious issues. A contentious issue is a subject that is likely to stimulate divergent viewpoints within people (e.g., Healthcare Reform, Same-Sex Marriage, Israel/Palestine conflict). In most cases opinion itself is not enough; arguments are needed when people differ on a specific issue. Multiple documents such as surveys' reports, debate sites' posts and editorials may contain multiple contrastive viewpoints regarding a particular issue of contention. Table 1 presents an example of short-text documents expressing divergent opinions where each is exclusively supporting or opposing a healthcare legislation[1]. Opinion in contentious issues is often expressed implicitly, not necessarily through the usage of usual negative or positive opinion words, like "bad" or "great". In addition, the propositional content of the utterances may remain ambiguous in certain circumstances. This makes its extraction a challenging task. Opinion is usually conveyed through the arguing expression justifying the endorsement of a particular point of view. It is advised by the stated words or phrases as they appear in the context. For example, the arguing expression "many people do not have healthcare", in Table 1,

---

[1] Extracted from a Gallup Inc. survey http://www.gallup.com/poll/126521/
favor-oppose-obama-healthcare-plan.aspx

E. Métais, M. Roche, and M. Teisseire (Eds.): NLDB 2014, LNCS 8455, pp. 114–125, 2014.
© Springer International Publishing Switzerland 2014

Table 1. Excerpts of support and opposition opinion to a healthcare bill in the USA

Support Viewpoint	Oppose Viewpoint
Many people do not have health care	The government should not be involved
Provide health care for 30 million people	It will produce too much debt
The government should help old people	The bill would not help the people

implicitly explains that the reform is intended to fix the problem of uninsured people, and thus, the opinion is probably on the supporting side. On the other hand, the arguing expression "it will produce too much debt" denotes the negative consequence that may result from passing the bill, making it on the opposing side.

Instead of going through the detailed contents of all documents provided by social media platforms, an automatic concise summary would be appealing for a number of users. For example, it may constitute a rich source of information for policy makers to monitor public opinion and feedback. For journalists, a substantial amount of work can be saved by having automatic access to drafting elements about controversial issues.

The rest of this paper is organized as follows. Section 2 states the problem. Section 3 explains the key issues in the context of recent related work. Section 4 provides the technical details of our model, the Joint Topic Viewpoint model (JTV). Section 5 describes the clustering task that might be used to obtain a feasible solution. Section 6 provides a description of the experimental set up on three different types of contentious text. Section 7 assesses the adequacy and compares the performance of our solution with another model in the literature. Section 8 concludes the paper.

2   Problem Statement

This paper introduces a method of mining important arguing expressions in different types of contentious text (surveys' reports, debate forums' posts and editorials). Table 2 presents an example of a human-made summary of arguing expressions [7], obtained from verbatim responses of a survey on the Obama healthcare. Given a corpus of documents, our ultimate goal is to generate similar summaries. However, this paper only concentrates on the subtask of mining the content by first identifying recurrent words and phrases expressing "arguments" and then clustering them according to their topics and viewpoints. Table 2's examples serve to define key concepts and help formulate the problem. Here, the contentious issue spawning the contrastive viewpoints is the Obama healthcare. The documents are people's verbatim responses to the question "Why do you favor or oppose a healthcare legislation similar to President Obama's ?".

We define a *contention question* as a question that can generate expressions of two or more divergent viewpoints as a response.

While the previous question explicitly asks for the reasons ("why"), we relax this constraint and consider also usual opinion questions like "Do you favor or oppose Obamacare ?", or "What do you think about Obamacare".

A *contentious document* is a document that contains expressions of one or more divergent viewpoints in response to the contention question.

Table 2. Human-made summary of arguing expressions supporting and opposing Obamacare

Support Viewpoint	Oppose Viewpoint
People need health insurance/many uninsured	Will raise cost of insurance/ less affordable
System is broken/needs to be fixed	Does not address real problems
Costs are out of control/help control costs	Need more information on how it works
Moral responsibility to provide/Fair	Against big government involvement (general)
Would make healthcare more affordable	Government should not be involved in healthcare
Don't trust insurance companies	Cost the government too much

Table 2 is split into two parts according to the viewpoint: supporting or opposing the healthcare bill. Each row contains one or more phrases, each expressing a reason (or an explanation), e.g., "System is broken" and "needs to be fixed". Though lexically different, these phrases share a common hidden theme (or topic), e.g., healthcare system, and implicitly convey the same hidden viewpoint's semantics, e.g., support the healthcare bill. Thus, we define an ***arguing expression*** as the set of reasons (snippets: words or phrases) sharing a common topic and justifying the same viewpoint regarding a contentious issue.

We assume that a ***viewpoint*** (e.g., a column of Table 2) in a contentious document is a stance, in response to a contention question, which is implicitly expressed by a set of arguing expressions (e.g., rows of a column in Table 2).

Thus, the arguing expressions voicing the same viewpoint differ in their topics, but agree in the stance. For example, arguing expressions represented by "system is broken" and "costs are out of control" discuss different topics, i.e., healthcare system and insurance's cost, but both support the healthcare bill. On the other hand, arguing expressions of divergent viewpoints may have similar topic or may not. For instance, "government should help elderly" and "government should not be involved" share the same topic "government's role" while conveying opposed viewpoints.

Our research problem and objectives in terms of the newly introduced concepts are stated as follows. Given a corpus of unlabeled contentious documents $\{doc_1, doc_2, .., doc_D\}$, where each document doc_d expresses one or more viewpoints v^d from a set of L possible viewpoints $\{v_1, v_2, .., v_L\}$, and each viewpoint v_l can be conveyed using one or more arguing expressions ϕ_l from a set of possible arguing expressions discussing K different topics $\{\phi_{1l}, \phi_{2l}, .., \phi_{Kl}\}$, the objective is to perform the following two tasks:

1. automatically extracting coherent words and phrases describing any distinct arguing expression ϕ_{kl};
2. grouping extracted distinct arguing expressions ϕ_{kl} for different topics, $k = 1..K$, into their corresponding viewpoint v_l.

This paper concentrates on the first task while discussing key elements to realize the second. For the first task, there is a need to account for arguing expressions related to the same topic and viewpoint but having different lexical features, e.g., "provide health care for 30 million people" and "many people do not have healthcare". For this purpose we propose a Joint Topic Viewpoint Model (JTV) to represent the mutual dependence between topics and viewpoints.

3 Related Work

Classifying Stances: An early body of work addresses the challenge of classifying viewpoints in contentious or ideological discourses using supervised techniques [8,10]. Although the models give good performance, they remain data-dependent and costly to label, making the unsupervised approach more appropriate for the existing huge quantity of online data. A similar trend of studies scrutinizes the discourse aspect of a document in order to identify opposed stances [12,16]. However, these methods utilize polarity lexicon to detect opinionated text and do not look for arguing expression, which is shown to be useful in recognizing opposed stances [14]. Somasundaran and Wiebe [14] classify ideological stances in online debates using generated arguing clues from the Multi Perspective Question Answering (MPQA) opinion corpus[2]. Our problem is not to classify documents, but to recognize recurrent pattern of arguing phrases instead of arguing clues. Moreover, our approach is independent of any annotated corpora.

Topic Modeling in Reviews Data: Another emerging body of work applies probabilistic topic models on reviews data to extract appraisal aspects and the corresponding specific sentiment lexicon. These kinds of models are usually referred to as joint sentiment/aspect topic models [6,17,18]. Lin and He [9] propose the Joint Sentiment Topic Model (JST) to model the dependency between sentiment and topics. They make the assumption that topics discussed on a review are conditioned on sentiment polarity. Reversely, our JTV model assumes that a viewpoint endorsement (e.g., oppose reform) is conditioned on the discussed topic (e.g., government's role). Moreover, JTV's application is different from that of JST. Most of the joint aspect sentiment topic models are either semi-supervised or weakly supervised using sentiment polarity words (Paradigm lists) to boost their efficiency. In our case, viewpoints are often expressed implicitly and finding specific arguing lexicon for different stances is a challenging task in itself. Indeed, our model is enclosed in another body of work based on a Topic Model framework to mine divergent viewpoints.

Topic Modeling in Contentious Text: A recent study by Mukherjee and Liu [11] examines mining contention from discussion forums data where the interaction between different authors is pivotal. It attempts to jointly discover contention/agreement indicators (CA-Expressions) and topics using three different Joint Topic Expressions Models (JTE). The JTEs' output is used to discover points (topics) of contention. The model supposes that people express agreement or disagreement through CA-expressions. However, this is not often the case when people express their viewpoint via other channels than discussion forums like debate sites or editorials. Moreover, agreement or disagreement may also be conveyed implicitly through arguing expressions rejecting or supporting another opinion. JTEs do not model viewpoints and use the supervised Maximum Entropy model to detect CA-expressions.

Recently, Gottipati et al. [3] propose a topic model to infer human interpretable text in the domain of issues using Debatepedia[3] as a corpus of evidence. Debatepedia is an online authored encyclopedia to summarize and organize the main arguments of two possible positions. The model takes advantage of the hierarchical structure of arguments

---

[2] http://mpqa.cs.pitt.edu/
[3] http://dbp.idebate.org

in Debatepedia. Our work aims to model unstructured online data, with unrestricted number of positions, in order to, ultimately, help extract a relevant contention summary.

The closest work to ours is the one presented by Paul et al. [13]. It introduces the problem of contrastive summarization which is very similar to our stated problem in Section 2. They propose the Topic Aspect Model (TAM) and use the output distributions to compute similarities' scores for sentences. Scored sentences are used in a modified Random Walk algorithm to generate the summary. The assumption of TAM is that any word in the document can exclusively belong to a topic (e.g., government), a viewpoint (e.g., good), both (e.g., involvement) or neither (e.g., think). However, according to TAM's generative model, an author would choose his viewpoint and the topic to talk about independently. Our JTV encodes the dependency between topics and viewpoints.

4 Joint Topic Viewpoint Model

Latent Dirichlet Allocation (LDA) [2] is one of the most popular topic models used to mine large text data sets. It models a document as a mixture of topics where each topic is a distribution over words. However, it fails to model more complex structures of texts like contention where viewpoints are hidden. We augment LDA to model a contentious document as a pair of dependent mixtures: a mixture of arguing topics and a mixture of viewpoints for each topic. The assumption is that a document discusses the topics in proportions, (e.g., 80% government's role, 20% insurance's cost). Moreover, as explained in Section 2, each one of these topics can be shared by divergent arguing expressions conveying different viewpoints. We suppose that for each discussed topic in the document, the viewpoints are expressed in proportions. For instance, 70% of the document's text discussing the government's role expresses an opposing viewpoint to the reform while 30% of it conveys a supporting viewpoint. Thus, each term in a document is assigned a pair topic-viewpoint label (e.g., "government's role-oppose reform"). A term is a word or a phrase i.e., n-grams ($n > 1$). For each topic-viewpoint pair, the model generates a topic-viewpoint probability distribution over terms. This topic-viewpoint distribution would correspond to what we define as an arguing expression in Section 2, i.e., a set of terms sharing a common topic and justifying the same viewpoint regarding a contentious issue.

Formally, assume that a corpus contains D documents $d_{1..D}$, where each document is a term's vector $\boldsymbol{w}_d$ of size N_d; each term w_{dn} in a document belongs to the corpus vocabulary of distinct terms of size V. Let K be the total number of topics and L be the total number of viewpoints. Let θ_d denote the probabilities (proportions) of K topics under a document d; ψ_{dk} be the probability distributions (proportions) of L viewpoints for a topic k in the document d (the number of viewpoints L is the same for all topics); and ϕ_{kl} be the multinomial probability distribution over terms associated with a topic k and a viewpoint l. The generative process (see. the JTV graphical model in Fig. 1) is:

- for each topic k and viewpoint l, draw a multinomial distribution over the vocabulary V: $\phi_{kl} \sim Dir(\beta)$;
- for each document d,
  - draw a topic mixture $\theta_d \sim Dir(\alpha)$
  - for each topic k, draw a viewpoint mixture $\psi_{dk} \sim Dir(\gamma)$

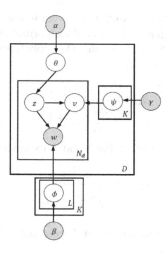

Fig. 1. The JTV's graphical model (plate notation)

- for each term w_{dn}
  - ∗ sample a topic assignment $z_{dn} \sim Mult(\theta_d)$
  - ∗ sample a viewpoint assignment $v_{dn} \sim Mult(\psi_{dz_{dn}})$
  - ∗ sample a term $w_{dn} \sim Mult(\phi_{z_{dn}v_{dn}})$

We use fixed symmetric Dirichlet's parameters γ, β and α. They can be interpreted as the prior counts of: terms assigned to viewpoint l and topic k in a document; a particular term w assigned to topic k and viewpoint l within the corpus; terms assigned to a topic k in a document, respectively. In order to learn the hidden JTV's parameters ϕ_{kl}, ψ_{dk} and θ_d, we draw on approximate inference as exact inference is intractable [2]. We use the collapsed Gibbs Sampling [4], a Markov Chain Monte Carlo algorithm. The collapsed Gibbs sampler integrate out all parameters ϕ, ψ and θ in the joint distribution of the model and converge to a stationary posterior distribution over viewpoints' assignments v and all topics' assignments z in the corpus. It iterates on each current observed token w_i and samples each corresponding v_i and z_i given all the previous sampled assignments in the model $v_{\neg i}$, $z_{\neg i}$ and observed $w_{\neg i}$, where $v = \{v_i, v_{\neg i}\}$, $z = \{z_i, z_{\neg i}\}$, and $w = \{w_i, w_{\neg i}\}$. The derived sampling equation is:

$$p(z_i = k, v_i = l | z_{\neg i}, v_{\neg i}, w_i = t, w_{\neg i}) \propto$$

$$\frac{n_{kl,\neg i}^{(t)} + \beta}{\sum\limits_{t=1}^{V} n_{kl,\neg i}^{(t)} + V\beta} \cdot \frac{n_{dk,\neg i}^{(l)} + \gamma}{\sum\limits_{l=1}^{L} n_{dk,\neg i}^{(l)} + L\gamma} \cdot n_{d,\neg i}^{(k)} + \alpha \quad (1)$$

where $n_{kl,\neg i}^{(t)}$ is the number of times term t was assigned to topic k and the viewpoint l in the corpus; $n_{dk,\neg i}^{(l)}$ is the number of times viewpoint l of topic k was observed in document d; and $n_{d,\neg i}^{(k)}$ is the number of times topic k was observed in document d.

All these counts are computed excluding the current token i, which is indicated by the symbol $\neg i$. After the convergence of the Gibbs algorithm, the parameters ϕ, ψ and θ are estimated using the last obtained sample. The probability that a term t belongs to a viewpoint l of topic k is approximated by:

$$\phi_{klt} = \frac{n_{kl}^{(t)} + \beta}{\sum\limits_{t=1}^{V} n_{kl}^{(t)} + V\beta}. \tag{2}$$

The probability of a viewpoint l of a topic k under document d is estimated by:

$$\psi_{dkl} = \frac{n_{dk}^{(l)} + \gamma}{\sum\limits_{l=1}^{L} n_{dk}^{(l)} + L\gamma}. \tag{3}$$

The probability of a topic k under document d is estimated by:

$$\theta_{dk} = \frac{n_{d}^{(k)} + \alpha}{\sum\limits_{k=1}^{K} n_{d}^{(k)} + K\alpha}. \tag{4}$$

5   Clustering Arguing Expressions

Although we are not tackling the task of clustering arguing expressions according to their viewpoints in this paper (Task 2 in Section 2), we explain how the structure of JTV lays the ground for performing it. We mentioned in the previous Section that an inferred topic-viewpoint distribution ϕ_{kl} can be assimilated to an arguing expression. For convenience, we will use "arguing expression" and "topic-viewpoint" interchangeably to refer to the topic-viewpoint distribution. Indeed, two topic-viewpoint ϕ_{kl} and $\phi_{k'l}$, having different topics k and k', do not necessarily express the same viewpoint, despite the fact that they both have the same index l. The reason stems from the nested structure of the model, where the generation of the viewpoint assignments for a particular topic k is completely independent from that of topic k'. In other words, the model does not trace and match the viewpoint labeling along different topics. Nevertheless, the JTV can still help overcome this problem. According to the JTV's structure, a topic-viewpoint ϕ_{kl}, is more similar in distribution to a divergent topic-viewpoint $\phi_{kl'}$, related to the same topic k, than to any other topic-viewpoint $\phi_{k'*}$, corresponding to a different topic k'. Therefore, we can formulate the problem of clustering arguments as a constrained clustering problem [1]. The goal is to group the similar topics-viewpoints ϕ_{kl}s into L clusters (number of viewpoints), given the constraint that the ϕ_{kl}s of the same topic k should not belong to the same cluster.

6   Experimental Set Up

In order to evaluate the performances of the JTV model, we utilize three types of multiple contrastive viewpoint text data: (1) short-text data where people express their viewpoint briefly with few words like survey's verbatim response or social media posts; (2)

Table 3. Statistics on the three used data sets

	GM		IP		OC	
Viewpoint	hurt	no	pal	is	for	ag
#doc	149	301	149	149	434	508
total #toks	47915		209481		14594	
avg. #toks per doc	106.47		702.95		15.94	

mid-range text where people develop their opinion further using few sentences, usually showcasing illustrative examples justifying their stances; (3) long text data, mainly
editorials where opinion is expressed in structured and verbose manner.

Throughout the evaluation procedure, analysis is performed on three different types
of data sets, corresponding to three different contention issues. Table 3 describes the
used data sets. **ObamaCare (OC)**[4] consists of short verbatim responses concerning
the "Obamacare" bill. The survey was conducted by Gallup®from March 4-7, 2010.
People were asked why they would oppose or support a bill similar to Obamacare. Table
2 is a human-made summary of this corpus. **Gay Marriage (GM)**[5] contains posts in
"createdebate.com" responding to the contention question "How can gay marriage hurt
anyone?". Users indicate the stance of their posts (i.e., "hurts everyone? (does hurt)"
or "doesn't hurt"). **Israel-Palestine (IP)**[6] data set is extracted from BitterLemons web
site. It contains articles of two permanent editors, a Palestinian and an Israeli, about
the same issue. Articles are published weekly from 2001 to 2005. They discuss several
contention issues, e.g., "the American role in the region" and "the Palestinian election".

Paul et al. [13] stress out the importance of negation features in detecting contrastive
viewpoints. Thus, we performed a simple treatment of merging any negation indicators,
like "nothing", "no one", "never", etc., found in text with the following occurring word
to form a single token. Moreover, we merge the negation "not" with any auxiliary verb
(e.g., is, was, could, will) preceding it. Then, we removed the stop-words.

Throughout the experiments below, the JTV's hyperparameters are set to fixed values. The γ is set, according to Steyvers and Griffiths's [15] hyperparameters settings, to
$50/L$, where L is the number of viewpoints. β and α are adjusted manually, to give reasonable results, and are both set to 0.01. Along the experiments, we try different number
of topics K. The number of viewpoints L is equal to 2. The TAM model [13] (Section
3) is run as a means of comparison during the evaluation (with default parameters).

7 Model Evaluation

7.1 Qualitative Evaluation

We perform a qualitative analysis of JTV using the ObamaCare data set. Tables 4
presents the inferred topic-viewpoints, i.e., arguing expressions. We set a number of

---

[4] http://www.gallup.com/poll/126521/
favor-oppose-obama-healthcare-plan.aspx
[5] http://www.createdebate.com/debate/show/
How_can_gay_marriage_hurt_any_one
[6] http://www.bitterlemons.net/

Table 4. JTV's generated topics-viewpoints from Obamacare data set

Topic 1 0.19		Topic 2 0.20		Topic 3 0.20	
View 1 0.55	View 2 0.45	View 3 0.51	View 4 0.49	View 5 0.54	View 6 0.46
healthcare	dont_think	people	government	insurance	country
system	work	cant_afford	dont_want	health	economy
uninsured	bill	doctors	involved	companies	medicine
country	abortion	lack	control	years	dollars
world	fair	covered	dont_think	prices	american
change	debt	americans	dont_like	reason	start

Topic 4 0.21		Topic 5 0.20	
View 7 0.55	View 8 0.45	View 9 0.47	View 10 0.53
healthcare	healthcare	people	costs
cost	cost	money	medicare
expensive	coverage	pay	increase
afford	dont_know	dont_have	pay
care	public	children	worse
feel	preexisting	poor	problems

topics of $K = 5$ and a number of viewpoints of $L = 2$. Each topic-viewpoint (e.g., Topic 1-View 1) is represented by the set of top terms. The terms are sorted in descending order according to their probabilities. Inferred probabilities over topics, and over viewpoints for each topic, are also reported. We try to qualitatively observe the distinctiveness of each arguing (topic-viewpoint) and assess its coherence in terms of the topic discussed and the viewpoint conveyed and its divergence with the opposing pair-element. In Table 4 most of the topic-viewpoint pairs, corresponding to the same topic, are conveying opposite stances. For instance, taking a closer look at the original data suggests that Topic 1-View 1 (Table 4) criticizes the healthcare system and stresses out the need for a change (e.g., " *We ought to **change** the **system** so everyone can have it (the healthcare insurance)*", "*Because the greatest **country** in the **world** has a dismal **healthcare system***"). This may correspond to the second support arguing expressions in the reference summary of Table 2. On the other side, Topic 1-View 2 may convey the belief that the bill will not work or that it is not fair e.g., "I **don't think** it's **fair**". It also opposes the bill for including the abortion and for the debt that it may induce. Although the debt and the abortion are not related, as topics, they both tend to be adduced by people opposing the bill. Similarly, Topic 2-View 3 may reveal that people can't afford healthcare and they need to be covered (first support arguing in Table 2). However, the opposite side seems to be not enthusiastic about the government's involvement and control (fourth and fifth oppose arguing expressions in Table 2). The same pattern is observed in Topic 3. A matching can also be established with the reference summary.

Detecting different arguing expressions for the same topic proves to be a difficult task when the reasons are lexically very similar. An example is Topic 4 in Table 4 where the shared topic is "healthcare cost". In this case, both arguing expressions are about high costs. The original data contains two rhetoric: one is about current existing costs (supporting side) and the other is about costs induced by the bill (opposing side). However, both Topic 4-View 7 and Topic 4-View 8 seem to convey the supporting viewpoint. The increasing costs yield by the bill may be conveyed in Topic 5-View 10.

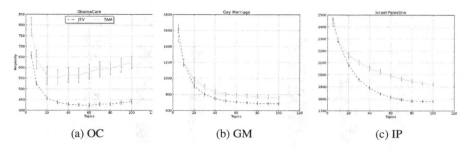

(a) OC   (b) GM   (c) IP

Fig. 2. JVT and TAM's perplexity plots for three different data sets

7.2   Quantitative Evaluation

We assess the ability of the model to fit three data sets and to generate distinct topic-viewpoint by comparing it with TAM which also models the topic-viewpoint dimension.

Held-Out Perplexity. We use the perplexity criterion to measure the ability of the learned topic model to fit a new held-out data. Perplexity assesses the generalization performance and, subsequently, provides a comparing framework of learned topic models. The lower the perplexity, the less "perplexed" is the model by unseen data and the better the generalization. It algebraically corresponds to the inverse geometrical mean of the test corpus' terms likelihoods given the learned model parameters [5]. We compute the perplexity under estimated parameters of JTV and compare it to that of TAM for our three unigrams data sets (Section 6). Figure 2 exhibits, for each corpus, the perplexity plot as function of the number of topics K for JTV and TAM. Note that for each K, we run the model 50 times. The drawn perplexity corresponds to the average perplexity on the 50 runs where each run compute one-fold perplexity from a 10-fold cross-validation. The figures show evidence that the JTV outperforms TAM for all data sets, used in the experimentation.

Kullback-Leibler Divergence. Kullback-Leibler (KL) Divergence is used to measure the degree of separation between two probability distributions. We utilize it for two purposes. The first purpose is to validate the assumption we made in Section 5 which states that, according to JTV's structure, a topic-viewpoint ϕ_{kl} is more similar in distribution to a topic-viewpoint $\phi_{kl'}$, related to the same topic k, than to any other topic-viewpoint $\phi_{k'*}$, corresponding to a different topic k'. Thus, two measures of *intra* and *inter-divergence* are computed. The *intra-divergence* is an average KL-Divergence between all topic-viewpoint distributions that are associated with a same topic. The *inter-divergence* is an average KL-Divergence between all pairs of topic-viewpoint distributions belonging to different topics. Figure 3a displays the histograms of JTV's intra and inter divergence values for the three data sets. These quantities are averages on 20 runs of the model for an input number of topics $K = 5$, which gives the best differences between the two measures. We observe that a higher divergence is recorded between

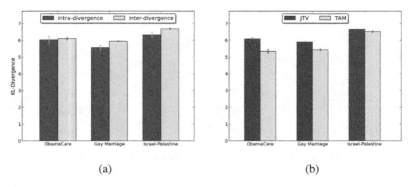

(a)   (b)

Fig. 3. Histograms of: (a) average topic-viewpoint intra/inter divergences of JTV; (b) average of overall topic-viewpoint divergences of JTV and TAM ($K = 5$)

topic-viewpoints of different topics than between those of a same topic. This is verified for all the data sets considered in our experimentation. The differences between the intra and inter divergences are significant ($p - value < 0.01$) over unpaired t-test (except for Obamacare). The second purpose of using KL-Divergence is to assess the distinctiveness of generated topic-viewpoint by JTV and TAM. This is an indicator of a good aggregation of arguing expressions. We compute an *overall-divergence* quantity, which is an average KL-Divergence between all pairs of topic-viewpoint distributions, for JTV and TAM and compare them. Figure 3b illustrates the results for all datasets. Quantities are averages on 20 runs of the models. Both models are run with a number of topics $K = 5$, which gives the best divergences for TAM. Comparing JTV and TAM, we notice that the overall-divergence of JTV's topic-viewpoint is significantly ($p - value < 0.01$) higher for all data sets. This result reveals a better quality of our JTV extracting process of arguing expressions (the first task stated in Section 2).

8   Conclusion and Future Work

Within the framework of probabilistic graphical models, we presented an approach to mining the important topics and divergent viewpoints in contentious opinionated text. We proposed a Joint Topic Viewpoint model (JTV) for the unsupervised detection and clustering of recurrent arguing expressions. Preliminary results show that our model can provide accommodation for various types of texts (survey reports, debate forums posts and editorials). Moreover, the detection and clustering accuracy has been shown to be enhanced by accounting for mutual dependence of topics and viewpoints. Future work study needs to improve the topicality coherence of extracted arguing phrases. It should also give more insights into their clustering according to their viewpoints, as well as their automatic extractive summary. A human-oriented evaluation of generated arguing expressions and summaries needs to be set up.

References

1. Basu, S., Davidson, I., Wagstaff, K.: Constrained Clustering: Advances in Algorithms, Theory, and Applications, 1st edn. Chapman & Hall/CRC (2008)
2. Blei, D.M., Ng, A.Y., Jordan, M.I.: Latent dirichlet allocation. Journal of Machine Learning Research 3, 993–1022 (2003)
3. Gottipati, S., Qiu, M., Sim, Y., Jiang, J., Smith, N.A.: Learning topics and positions from debatepedia. In: Proceedings of Conference on Empirical Methods in Natural Language Processing (2013)
4. Griffiths, T.L., Steyvers, M.: Finding scientific topics. Proceedings of the National Academy of Sciences of the United States of America 101(1), 5228–5235 (2004)
5. Heinrich, G.: Parameter estimation for text analysis. Tech. rep., Fraunhofer IGD (September 2009)
6. Jo, Y., Oh, A.H.: Aspect and sentiment unification model for online review analysis. In: Proceedings of the Fourth ACM International Conference on Web Search and Data Mining, pp. 815–824 (2011)
7. Jones, J.M.: In u.s., 45% favor, 48% oppose obama healthcare plan (March 2010), http://www.gallup.com/poll/126521/favor-oppose-obama-healthcare-plan.aspx
8. Kim, S.M., Hovy, E.H.: Crystal: Analyzing predictive opinions on the web. In: Joint Conference on Empirical Methods in Natural Language Processing and Computational Natural Language Learning, pp. 1056–1064 (2007)
9. Lin, C., He, Y.: Joint sentiment/topic model for sentiment analysis. In: Proceedings of the 18th ACM Conference on Information and Knowledge Management, pp. 375–384 (2009)
10. Lin, W.H., Wilson, T., Wiebe, J., Hauptmann, A.: Which side are you on?: Identifying perspectives at the document and sentence levels. In: Proceedings of the Tenth Conference on Computational Natural Language Learning, pp. 109–116 (2006)
11. Mukherjee, A., Liu, B.: Mining contentions from discussions and debates. In: Proceedings of the 18th ACM SIGKDD International Conference on Knowledge Discovery and Data Mining, pp. 841–849 (2012)
12. Park, S., Lee, K., Song, J.: Contrasting opposing views of news articles on contentious issues. In: Proceedings of the 49th Annual Meeting of the Association for Computational Linguistics: Human Language Technologies, pp. 340–349 (2011)
13. Paul, M.J., Zhai, C., Girju, R.: Summarizing contrastive viewpoints in opinionated text. In: Proceedings of the 2010 Conference on Empirical Methods in Natural Language Processing, pp. 66–76 (2010)
14. Somasundaran, S., Wiebe, J.: Recognizing stances in ideological on-line debates. In: Proceedings of the NAACL HLT 2010 Workshop on Computational Approaches to Analysis and Generation of Emotion in Text, pp. 116–124 (2010)
15. Steyvers, M., Griffiths, T.: Probabilistic topic models. In: Handbook of Latent Semantic Analysis, vol. 427(7), pp. 424–440 (2007)
16. Thomas, M., Pang, B., Lee, L.: Get out the vote: Determining support or opposition from congressional floor-debate transcripts. In: Proceedings of the 2006 Conference on Empirical Methods in Natural Language Processing, pp. 327–335 (2006)
17. Titov, I., McDonald, R.: Modeling online reviews with multi-grain topic models. In: Proceedings of the 17th International Conference on World Wide Web, pp. 111–120 (2008)
18. Zhao, W.X., Jiang, J., Yan, H., Li, X.: Jointly modeling aspects and opinions with a maxent-lda hybrid. In: Proceedings of the 2010 Conference on Empirical Methods in Natural Language Processing, pp. 56–65 (2010)

Identification of Multi-Focal Questions
in Question and Answer Reports

Mona Mohamed Zaki Ali[1,2], Goran Nenadic[3], and Babis Theodoulidis[4]

[1] School of Computer Science, The University of Manchester, Manchester, United Kingdom
[2] Faculty of Computers and Informatics, Suez Canal University, Ismailia, Egypt
mohamem@cs.man.ac.uk
[3] School of Computer Science, The University of Manchester, Manchester, United Kingdom
g.nenadic@manchester.ac.uk
[4] Manchester Business School, The University of Manchester, Manchester, United Kingdom
b.theodoulidis@manchester.ac.uk

Abstract. A significant amount of business and scientific data is collected via question and answer reports. However, these reports often suffer from various data quality issues. In many cases, questionnaires contain a number of questions that require multiple answers, which we argue can be a potential source of problems that may lead to poor-quality answers. This paper introduces multi-focal questions and proposes a model for identifying them. The model consists of three phases: question pre-processing, feature engineering and question classification. We use six types of features: lexical/surface features, Part-of-Speech, readability, question structure, wording and placement features, question response type and format features and question focus. A comparative study of three different machine learning algorithms (Bayes Net, Decision Tree and Support Vector Machine) is performed on a dataset of 150 questions obtained from the Carbon Disclosure Project, achieving the accuracy of 91%.

Keywords: Question Classification, Question Analysis, Content Analysis, Data Quality, Text Mining, Data Mining, Machine Learning, Rule-based Methods.

1   Introduction

Most business and scientific data are represented in unstructured or semi-structured formats. Some estimates indicate that more than 85% of business information exists as unstructured data [1]. In particular, Question and Answer (Q&A) reports are gaining momentum as a way to collect responses that can be used by data brokers and analysts (e.g., customer satisfaction reports, FAQ and sustainability reports). A Q&A report typically contains a series of questions that can be answered by providing either a structured answer from a pre-defined set of values (e.g. multiple choice test answering, drop down menus, multi-selection checkboxes, etc.) or by unstructured (typically textual) responses written in natural language.

Typically, a questionnaire is a useful data collection method that can yield high quality data if designed carefully [2]. Ideally, a questionnaire should comprise questions

E. Métais, M. Roche, and M. Teisseire (Eds.): NLDB 2014, LNCS 8455, pp. 126–137, 2014.

that are "clear, unambiguous, and understandable" [3] and, in particular, have each question asking for only one piece of information at a time [4]. However, in practice, questions are often complex and ask for several facts or answers to be provided. We argue that this may lead to data providers supplying partial answers, e.g. by forgetting or ignoring requests for multiple pieces of information, which consequently results in low-quality data. Identifying questions that ask for multiple answers may help Q&A designers in engineering data collections that will reduce the possibility for entering low quality data.

In this paper we introduce multi-focal questions (MFQs) as questions that request more than one unit of information, where a unit of information focuses on or links to a single concept in the domain. We propose a model for identifying these via question classification using machine learning and rule-based classifiers.

The remainder of this paper is organised as follows: Section 2 briefly reviews previous work on question classification methods. Section 3 explains the methods: we briefly describe the data set in 3.1, and present feature engineering and extraction are in 3.2 and 3.3. The question classification model is presented in 3.4. The model evaluation, experimental results and discussion are presented in Section 4, while Section 5 concludes the paper and offers some ideas for future work.

2   Related Work

Question classification is considered a crucial process in question processing for various applications, mainly for Question Answering (QA) systems [5, 6]. In a typical QA system, a question classification process analyses a given question to determine its type; i.e., question asks for a person, location, product, etc.; or for definition, description, reason, etc. [6]; or determines the focus of the question [7]. In the context of QA systems evaluation, the answer type and cardinality were used to determine the number of expected answers to be returned for a set of questions using deep linguistic analysis of questions and answers [8]. In this paper, we introduce the expected answer type and/ or format as well as the question structure, wording and placement as part of the feature types in our methods.

Proposed approaches for question classification range from hand-crafted rules [13-14] to machine learning [15, 16]. Machine-learning techniques have proved to perform better and be more efficient than rules [5], [6], [15] since the latter require extensive effort [15], [17] and usually suffer from low coverage [6]. Machine-learning techniques have been used successfully in a number of benchmark problems and achieved high performance and more accurate question classification [15, 16]. Recent research results have also shown that using a combination of machine learning and rule-based techniques could combine the advantages of both [7].

Questionnaire design and development research has presented different types of question such as: nested questions [3], probing questions [9], decomposition (multiple) questions [10], conditional questions [11], redirecting questions [3], and structured and/ or matrix questions [12]. For the purposes of this study, we extended the definitions of some of these question types in our methods.

3   Method

Our aim is to build a machine learning classifier that will for a given question return whether or not it is MFQ; i.e. whether it requires more than one unit of information in the answer. For the purpose of this study, a question can appear in the form of an interrogative sentence; i.e. an explicit question, such as: "How does climate change present general opportunities for your company?" or a request to provide information, such as: *"Please state the methodology and data sources you have used for calculating these reductions and savings"*. An example of MFQ: *"If your goods and/or services enable GHG emissions to be avoided by a third party, please provide details including the estimated avoided emissions, the anticipated timescale over which the emissions are avoided and the methodology, assumptions, emission factors (including sources), and global warming potentials (including sources) used for your estimations"*. An example of not-MFQ: *"How is your company exposed to physical risks from climate change?"*

3.1   Data Set

To investigate and validate our model, we use the Carbon Disclosure Project (CDP) as a case study. It produces annual Q&A reports to be used by investment managers and brokers [18]. Over 4000 organisations across the globe now measure and disclose their greenhouse gas emissions and climate change strategies through the CDP [18]. Between 2008 and 2010, the average number of questions is 50 questions per questionnaire, and the average length of questions is 3 sentences per question.

For our study, we manually analysed a relatively large set of Q&A reports (600 reports for 2008-2010). A dataset of 150 questions (from 2008-2010) was manually annotated as being MFQ or not (81 MFQ; 69 not-MFQ). A gold standard set was generated, and Inter Annotator Agreement (IAA) estimated by Kappa [19] was 0.8 (8 annotators were involved) which indicated a strong human agreement.

3.2   Feature Engineering

We designed a machine learning based classifier using a set of features engineered from each question. Fig. 1 illustrates the feature engineering process. Each question is represented as a feature vector consisting of over 60 features divided into six types (see Table 1 for summary). Feature types include:

1. **Lexical/surface features** include the number of words, number of sentences, number of syllables in the question, average number of syllables/words in the question, and average number of words/sentences in the question.
2. **Part-of-Speech (POS) distribution features** include the distribution of POS and lemma information for each sentence per question.

3. **Readability feature** measures text difficulty and understanding level to measure the reading-ease levels of each question.
4. **Question structure, wording and placement features** include the physical/linguistic structure of the question, how the question is phrased and the ordering of the questions in the questionnaire. This relates to question *composition* (how the question is composed), *linkage* (whether the question refers to another) and *placement*. The corresponding features are:

- **Nested.** We introduce a nested question as one that consists of a number of other questions. It is a categorical feature with values of *'probing'*, *'multiple'*, *'both'* and *'neither'*. We specify a **probing** question as one that appears as part of another question in order to go under the surface of an initial answer; i.e. respondents are asked to provide details of their choice. It is often recognised by statements like *"please provide details"* or similar (e.g. *"please provide further information/attach a document or a link that illustrates more details"*. We use a **multiple** question to represent a question that consists of two or more questions, with each often requiring a separate answer. Questions can be: a) separated by a comma, question mark, full stop, etc. (e.g. *"What emissions reductions, energy savings and associated cost savings have been achieved to date as a result of the plan and/or the activities described above? Please state the methodology and data sources you have used for calculating these reductions and savings"*); or b) divided into one or more sub-questions (e.g. *"Does a Board Committee or other executive body have overall responsibility for climate change? If   not: Please state how overall responsibility for climate change (…). If so, please provide the following information: i. Which Board Committee or executive body has overall responsibility (…)? ii. What is the mechanism by which the Board (…)?"*).

- **Conditional.** This is a binary feature. We introduce a conditional question to represent a question that consists of one or more conditional instruction(s) or statement(s). It can appear single, as part of another question or a sub-question. Examples of conditional statement(s)/sentence(s) include: a) Dual or "if-else" conditional statement(s); "if yes/so" or "If not/no/other". This type of conditional statement follows responses that are almost entirely dichotomous (yes/no) in nature (e.g. *"Does a Board Committee (…)? If not, please state (…). If so, please provide (…)"*). b) Other forms of conditional sentence(s), such as "in case of…" (e.g. *"If you have an absolute target: Please provide (). If it is an intensity target: Please provide (…).For both types of target, also: Please provide (…)."*).

- **Related.** This is a binary feature. We introduce a related question as one that relates to a separate question by referring back or forward or by explicitly mentioning its number (e.g. *"Has any of the information reported in response to questions 10–15 been externally verified/assured (…)?"*).

- **Layout/Section**. This is a categorical feature illustrates of which section a given question is part (it is specific to questionnaire structure and domain knowledge). In the case of CDP, four possible values/ categories for this feature were extracted: *'strategy'; 'greenhouse gas emissions'; 'risks and opportunities';* and *'performance and governance'.*

5. **Question response (answer) type and/or format features** include the format or structure of the expected answer from respondents. The corresponding features are:

- **Open-ended.** An open-ended question is one that allows respondents to provide "their own response, in their own words" [20], and it "does not limit the response to a single type of answer" [3] (e.g. *"What are the main sources of uncertainty in your data gathering, handling and calculations; e.g. data gaps, assumptions, extrapolation, metering/measurement inaccuracies etc?"*). "Open-ended" is a categorical feature with values of *'open'*, *'fixed'* and *'semi-fixed'*.

- **Tabulated**. This is a binary feature. We introduce a tabulated question as one that instructs respondents to provide their answers using a table or grid, which is either pre-provided by the questionnaire designer (s) or they are asked to formulate a table or grid of their choice. In both cases, the answer should be provided within a tabulated structure (e.g. *"Please use the table to give the total amount of fuel (…). Please complete the table by breaking down (…)"*).

6. **Question focus feature** is a categorical feature illustrates the focus of the question. It is a domain specific feature; i.e. it requires domain knowledge and/or domain expert(s) to review and filter the extraction results. In the case of CDP, eight possible values/ categories for this feature were extracted: *'emissions and methodology'; 'direct and indirect scopes'; 'costs and measuring'; 'risks and opportunities'; 'strategy and communications;' data accuracy and verification'; 'targets and initiatives';* and *'reporting boundary'.*

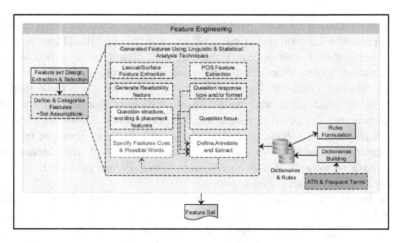

Fig. 1. Feature Engineering Process

Table 1. Summary of feature types, and their corresponding categories and values. Strings in *italic* represent feature values/categories.

Feature Type		Value type	Feature examples
Lexical/ Surface		Numeric	#words, #sentences, #syllables/Q, avg #syllables/words/Q, etc
POS distribution		Numeric	#NN/sentence, #VV/sentence, etc
Readability		Categorical	*Very difficult; difficult; fairly difficult; standard; fairly easy and easy*
Question structure, wording & placement	Nested	Categorical	*probing; multiple; both; neither*
	Conditional	Binary	*T,F*
	Related	Binary	*T,F*
	Layout/ Section	Categorical	*greenhouse gas emissions; risks and opportunities*, etc.
Question response type	Open-ended	Categorical	*open; fixed; semi-fixed*
	Tabulated	Binary	*T,F*
Question focus		Categorical	*emissions and methodology; direct and indirect scopes*, etc.

3.3   Feature Extraction

The feature extraction process differs for each feature type. For each question, POS tags were extracted using Tree Tagger [21] for every token for every sentence, and the distribution of each tag was calculated per sentence. The "Readability" feature was extracted using the Flesch formula [22, 23]. The resulting reading-ease levels were categorised in one of six categories: 'very difficult'; 'difficult'; 'fairly difficult'; 'standard'; 'fairly easy', and 'easy'. We used the boundaries of the ranges of each of these typically reported in the literature [24, 25].

The feature extraction process for the rest of features is illustrated in Fig. 2. For question structure, wording and placement features, question response type and/or format features and "Question focus" feature, the feature extraction process relies on rules and dictionaries. We created rules using the SPSS Modeler [26, 27] – SPSS has its own scripting language for rule generation - in combination with various types of dictionary - covering different categories such as: multi-word terms (~160 terms), single-word terms (~30 terms), named entity terms (~1600 terms; cover finance, location, date, status, organisations, etc.), general business terms (~110 terms), and domain specific terms (~80 terms), etc.

Initially, the dictionaries are built by extracting frequent terms using Automatic Term Recognition (ATR) tools, where candidate multi-word terms and their variants are recognised using a combination of linguistic filters and statistical analysis [28]. We use ATR results as seeds for SPSS Modeler libraries' developer [26, 27]. ATR results are imported and compiled in SPSS Modeler, and additional term extraction process is conducted within the Modeler to extract both 'single-word' and 'multi-word' terms. Within SPSS Modeler, the extraction process relies on linguistic-based text analysis which uses a set of linguistic resources in the form of: general dictionaries containing a list of base forms with a POS code, domain specific dictionaries (user defined), substitution dictionaries (contain synonyms) and exclude dictionaries [26, 27]. All types of dictionary are used to aid the identification of candidate terms which are words or groups of words used to identify relevant concepts in text. After candidate terms are identified, dictionaries are updated.

For each feature, we run a categorisation process for feature extraction (see Fig. 2). Each category corresponds to a specific feature value. For instance, eight possible categories (see 3.2) were extracted for the "Question focus" feature. Each category is identified using a set of descriptors. Descriptors are used to identify whether or not a question belongs to a specific category. They consist of candidate terms in dictionaries, and hand-crafted category and pattern matching rules [27]. Category labels are created automatically for "Layout/Section" feature and "Question focus" feature, with each category label (corresponding to a specific feature value) represents higher level concept(s) or topic(s) capturing the knowledge expressed in text. For "Nested" and "Open-ended" features, category labels are predefined; for instance, 'probing' for "Nested" feature and 'open' for "Open-ended" feature. For binary features, such as "Tabulated" feature, category labels take the same name as the feature type.

Rules are based on pattern matching; i.e., the text within each question is scanned to see whether any text matches a descriptor; if a match is found, this information is extracted as a pattern and the question is assigned to that category. The rules act as a Boolean query used to perform a match on a sentence. They contain one or more of the following arguments: types, macros, literal strings, or word gaps [26, 27]. Results are fine-tuned using linguistic techniques - co-occurrence and concept inclusion algorithms [26, 27] were used in our case.

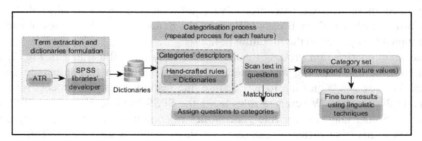

Fig. 2. Categorisation process for feature extraction; repeated process for each feature

3.4   MFQ Classification

MFQ classification is designed as a binary classification. A comparative study of three machine learning algorithms: Bayes Net (BN), Decision tree (DT) and Support Vector Machine (SVM) has been conducted for the same question-classification task in order to measure the stability of the classification accuracy and the performance of our model. Different types of kernels (Radial Basis Function (RBF), polynomial and linear kernels) were trained and tested for SVM. Default values for the parameters for each of the other learning algorithms were used. The previously presented feature set is automatically extracted for each question and used for each classifier.

K-fold cross-validation [29, 30] and repeated sub-sampling validation [30, 31] have been used to evaluate the accuracy of corresponding classifiers. An average accuracy rate for five conducted runs for each algorithm and per each split has been calculated and reported. The question classification performance is measured by accuracy, i.e. the proportion of the correctly classified questions among all test questions [15].

4   Results and Discussion

We trained and tested the methods on 150 questions. We acknowledge that some of the questions could be repeated in the same or similar form over years. As shown in tables 2 and 3, BN showed quite a stable accuracy of 88% and 89.3%. DT and SVM-polynomial kernel achieved the highest accuracy of 90.7% and 89% respectively. In addition, using partitioning of 60-40 for training and testing, both SVM-polynomial kernel and linear kernel achieved almost the same accuracy of 85.3% and 85.7% respectively. SVM high performance and stability match previous work results for different question classification tasks [6], [15].

Table 2. K-fold cross-validation: the average accuracy results for 5 runs for each algorithm and for each K

K	DT	BN	SVM		
			RBF	Polynomial	Linear
5	88.2	87.1	66.1	87.4	87.2
10	90.4	88.0	68.5	88.7	87.2
20	**90.7**	**88.0**	68.7	**89.0**	86.1

Table 3. Repeated sub-sampling validation: the average accuracy results for 5 runs for each algorithm for each split

Training-Testing	DT	BT	SVM		
			RBF	Polynomial	Linear
40-60	82.5	85.0	56.0	80.3	81.1
50-50	**86.4**	85.6	50.1	84.3	84.3
60-40	83.6	84.3	56.6	**85.3**	**85.7**
70-30	86.2	**89.3**	56.7	83.6	81.0

To test the influence of different features on the classification task, further evaluations and experimental comparisons were conducted using feature selection. A further experiment was conducted using filtering of the original feature set that excluded POS distribution features (these features were chosen for exclusion in order to test their effect on the performance as the number of extracted POS features is relatively high). Each classifier was tested and accuracy averages were calculated. A summary of the results together with the top 5 contributing features from feature selection (before and after excluding POS distribution features) are shown in tables 4 and 5. Table 5 shows that POS features tend to increase the performance constantly. Further, the "Nested" feature tend be the key feature in the classification, along with the "Conditional" feature.

Table 4. The average accuracy results for 5 runs for each algorithm using 20-fold cross validation, excluding POS distribution features from the feature set

Algorithm	Accuracy (%)	Top 5 contributing features in descending order (feature selection)
SVM	88.4	Nested, Conditional, #words, #sentences, Readability
BN	87.2	
DT	89.5	

Table 5. The average accuracy results for 5 runs for each algorithm using 20-fold cross validation, with the original feature set

Algorithm	Accuracy (%)	Top 5 contributing features in descending order (feature selection)
SVM	89.0	Nested, POS_NNS, POS_VV POS_CC, Conditional
BN	88.0	
DT	90.7	

Another interesting feature is "Readability"; Fig. 3 shows the distribution where the percentage of 'very difficult' and 'difficult' questions is relatively high (reaching more than 79% of the total number of our questions set). Also, it was observed that there is a positive relationship between the volume of domain concepts and expressions in text on one hand and the readability and text understanding score on the other.

Value	Proportion	%	Count
Very_Difficult		43.33	65
Difficult		36.0	54
Standard		9.33	14
Fairly_Difficult		8.0	12
Fairly_Easy		2.0	3
Easy		1.33	2

Fig. 3. Distribution of question readability and level of text understanding difficulty across 150 questions

Furthermore, association rules were used to explore the relationships between MFQ and all features. We used most confident association rules to build a rule-based classifier. The classifier was tested (on a testing set of 50% of our gold standard)

using the top rules uncovered by association rules generated from the Apriori algorithm (using the SPSS Modeler) [26]. It is interesting that the accuracy rate reached 90% using only the "Nested" feature (Table 6).

Table 6. Correlations between MFQ and feature set using association rules, and presenting the resulted rule-based classification accuracies using the presented participating features/ rules and 50% testing set

Participating feature(s) (Rules)	Confidence (%)	Rule-based Classifier Accuracy (%)
Nested=probing, Nested=multiple	100	90
Nested=probing, Nested=multiple, Open-ended=open	100	90
Conditional, Nested=multiple	100	90
Readability=Difficult, Nested=probing, Nested=multiple	100	90
Readability=Difficult, Nested= probing, Nested= multiple, Open-ended=open	100	76
Readability=Difficult, Nested=multiple	100	71
Nested=multiple	95.7	**90**
Nested=multiple, Open-ended=open	95.5	90
Readability=Difficult, Nested=probing	91.0	75
Conditional	91.0	76
Nested=probing, Open-ended=open	82.0	76
Nested=probing	78.3	74

5   Conclusion

This paper presents a model for identifying multifocal questions as these are one of the potential sources of DQ problems within Q&A reports through automatic question classification. We showed that we can recognise MFQ using a set of automatically extracted features with relatively good accuracy. Some of the extracted features are domain specific, such as "Question focus"; i.e. they require a corresponding domain knowledge and/or domain expert(s) to review and filter the extraction results. Association rules and rule-based classification demonstrated the "Nested" feature to be the strongest indicator of MFQ. Experimental results show that using rules can achieve a comparable performance to machine learning for MFQ identification.

The main future direction of our research is to exploit more/ in depth domain knowledge perspective for question classification. It is also worth investigating other types of machine learning algorithms as well as cascading and fuzzy classification. In addition, we plan to expand the question analysis to cover the "used medium" (the formatting of question(s) and the type and format of the provided answer space) as we hypothesise that it has a potential effect on the quality of data collected from Q&A reports.

Acknowledgments. This research is funded by the High Ministry of Education in Egypt. Also, we would like to acknowledge the ongoing support of the Egyptian Cultural and Educational Bureau in London, and The University of Manchester EPS Graduate and Researcher Development Committee.

References

1. Blumberg, R., Atre, S.: The problem with unstructured data. DM Review 13, 42–49 (2003)
2. Marshall, G.: The purpose, design and administration of a questionnaire for data collection. Radiography 11(2), 131–136 (2005)
3. Fadem, T.J.: The art of asking: ask better questions, get better answers. FT Press (2008)
4. Leung, W.-C.: How to design a questionnaire. BMJ 9(11), 187–189 (2001)
5. Huang, P., Bu, J., Chen, C., Qiu, G.: An effective feature-weighting model for question classification. In: Computational Intelligence and Security International Conference, pp. 32–36. IEEE (2007)
6. Tamura, A., Takamura, H., Okumura, M.: Classification of multiple-sentence questions. In: Dale, R., Wong, K.-F., Su, J., Kwong, O.Y. (eds.) IJCNLP 2005. LNCS (LNAI), vol. 3651, pp. 426–437. Springer, Heidelberg (2005)
7. Xiao-Ming, L., Li, L.: Question Classification Based on Focus. In: 2012 International Conference Communication Systems and Network Technologies (CSNT), pp. 512–516. IEEE (2012)
8. Bos, J.: The "La Sapienza" Question Answering System at TREC-2006. In: Voorhees, E.M., Buckland, L.P. (eds.) The Fifteenth Text RETrieval Conference, Gaitersburg, MD, pp. 797–803 (2006)
9. Sahin, A., Kulm, G.: Sixth grade mathematics teachers' intentions and use of probing, guiding, and factual questions. Journal of Mathematics Teacher Education 11(3), 221–241 (2008)
10. Hagstrom, P.A.: Decomposing questions. PhD dissertation, Massachusetts Institute of Technology (1998)
11. Isaacs, J., Rawlins, K.: Conditional questions. Journal of Semantics 25(3), 269–319 (2008)
12. Rubin, A., Babbie, E.R.: Research methods for social work. Cengage Learning (2008)
13. Voorhees, E.M.: Overview of the TREC 2001 question answering track. In: NIST Special Publication, pp. 42–51 (2002)
14. Sehgal, A.K., Das, S., Noto, K., Saier, M.K., Elkan, C.: Identifying relevant data for a biological database: Handcrafted rules versus machine learning. IEEE/ACM Transactions Computational Biology and Bioinformatics 8(3), 851–857 (2011)
15. Zhang, D., Lee, W.S.: Question classification using support vector machines. In: Proceedings of the 26th Annual International ACM SIGIR Conference on Research and Development in Information Retrieval, pp. 26–32. ACM (2003)
16. Loni, B., van Tulder, G., Wiggers, P., Tax, D.M.J., Loog, M.: Question classification by weighted combination of lexical, syntactic and semantic features. In: Habernal, I., Matoušek, V. (eds.) TSD 2011. LNCS (LNAI), vol. 6836, pp. 243–250. Springer, Heidelberg (2011)
17. Metzler, D., Croft, W.B.: Analysis of statistical question classification for fact-based questions. Information Retrieval 8 3, 481–504 (2005)
18. Carbon Disclosure Project, https://www.cdproject.net
19. Artstein, R., Poesio, M.: Inter-coder agreement for computational linguistics. Computational Linguistics 34(4), 555–596 (2008)

20. Murray, P.: Fundamental issues in questionnaire design. Accident and Emergency Nursing 7(3), 148–153 (1999)
21. TreeTagger - a language independent part-of-speech tagger, http://www.cis.uni-muenchen.de/~schmid/tools/TreeTagger/
22. Flesch, R.: A new readability yardstick. Journal of Applied Psychology 32, 221 (1948)
23. Kincaid, J.P., Fishburne Jr., R.P., Rogers, R.L., Chissom, B.S.: Derivation of new readability formulas (automated readability index, fog count and flesch reading ease formula) for navy enlisted personnel. Naval Technical Training Command Millington TN Research Branch (1975)
24. Flesch Reading Ease Readability Score, http://rfptemplates.technologyevaluation.com/readability-scores/flesch-reading-ease-readability-score.html
25. Flesch, R.F.: How to test readability. Harper (1951)
26. IBM SPSS Modeler for data and text mining, http://www.01.ibm.com/software/analytics-/spss-/products/modeler/
27. IBM SPSS Modeler Text Analytics, ftp://public.dhe.ibm.com/software/analytics/spss/documentation/modeler/15.0/en/Users_Guide_For_Text_Analytics.pdf
28. Nenadié, G., Ananiadou, S., McNaught, J.: Enhancing automatic term recognition through recognition of variation. In: Proceedings of the 20th International Conference on Computational Linguistics, p. 604. ACL (2004)
29. Bishop, C.M., Nasrabadi, N.M.: Pattern recognition and machine learning, vol. 1. Springer, New York (2006)
30. Kantardzic, M.: Data mining: concepts, models, methods, and algorithms. John Wiley & Sons (2011)
31. Li, D.-C., Fang, Y.-H., Fang, Y.M.: The data complexity index to construct an efficient cross-validation method. Decision Support Systems 50(1), 93–102 (2010)

Improving Arabic Texts Morphological Disambiguation Using a Possibilistic Classifier

Raja Ayed[1], Ibrahim Bounhas[2], Bilel Elayeb[1,3], Narjès Bellamine Ben Saoud[1,4], and Fabrice Evrard[5]

[1] RIADI Research Laboratory, ENSI Manouba University 2010, Tunisia
ayed.raja@gmail.com, Bilel.Elayeb@riadi.rnu.tn
Narjes.Bellamine@ensi.rnu.tn
[2] LISI Lab. of Computer Science for Industrial Systems, ISD Manouba University, 2010 Tunisia
Bounhas.Ibrahim@yahoo.fr
[3] Emirates College of Technology, P.O. Box: 41009, Abu Dhabi, United Arab Emirates
[4] Higher Institute of Informatics (ISI), Tunis El Manar University, 1002 Tunisia
[5] Informatics Research Institute of Toulouse (IRIT), 02 Rue Camichel, 31071 Toulouse, France
Fabrice.Evrard@enseeiht.fr

Abstract. Morphological ambiguity is an important problem that has been studied through different approaches. We investigate, in this paper, some classification methods to disambiguate Arabic morphological features of non-vocalized texts. A possibilistic approach is improved and proposed to handle imperfect training and test datasets. We introduce a data transformation method to convert the imperfect dataset to a perfect one. We compare the disambiguation results of classification approaches to results given by the possibilistic classifier dealing with imperfection context.

Keywords: Arabic Morphological Disambiguation, Possibilistic Classification, Imprecise and Uncertain Datasets.

1 Introduction

Various words in Arabic language may have identical morphological form because of the lack of short vowels in some texts which present a challenge for Arabic Natural Language Processing (NLP) task [1]. Indeed, one non-vocalized word may have more than 10 interpretations. For instance, if we add short vowels to the form "كتب", we can obtain "كُتُب" (books) or "كَتَبَ" (he wrote) and so on. Many morphological analyzers provide the stem and the flectional marks of one Arabic word [2]. A morphological analyzer describes the different values of morphological features (i.e. part-of-speech, gender, number, voice, etc.) independently of the word context. It assigns to a given word its possible solutions. If the analyzer provides more than one solution for a given word, this word will be considered ambiguous. So, we need applying morphological disambiguation which consists in selecting the most accurate value among the proposed solutions. Hajic [2] confirmed that morphological disambiguation may be

E. Métais, M. Roche, and M. Teisseire (Eds.): NLDB 2014, LNCS 8455, pp. 138–147, 2014.

supported by a morphological analyzer. Many works use the classification approaches to resolve the morphological disambiguation task [3, 4, 5].

In this paper, we consider the morphological features as classes. We aim to run training phase on vocalized texts since they are less ambiguous. Hence, we disambiguate words of non-vocalized texts by matching their correct values of the morphological features. The training and test sets contain instances described by attributes. Each instance, from the training set, is assigned to a vocalized word w. This instance defines the morphological feature (the class) of this word w. The test set presents non-classified instances. Both training and test sets may include ambiguous instances. In other terms, these instances may provide more than one value of each classification attribute. These instances are considered imperfect (imprecise). Hence, the Arabic text disambiguation consists in classifying the different morphological features through a perfect or an imperfect training set [6, 7]. Thus, we study classification approaches to disambiguate non-vocalized texts using vocalized texts for training. We investigate the application and the efficacy of an original possibilistic classification process described in [6, 7] and we improve it by adding a data transformation step to deal with imprecise data. The possibilistic approach handles the imperfection case [9]. We compare classification results, given by various classifiers, with the possibilistic approach results.

This paper is organized as follows. Related classification approaches are briefly presented in Section 2. Section 3 details the proposed improvement of a possibilistic classifier. In Section 4, we expose the general architecture of the classification model. Experimental results and their comparative study are given in Section 5. Finally, we conclude our work and we propose some future extensions in Section 6.

2   Related Classifiers

Classifiers assign, from a predefined set, a class to an object depending on the values of attributes used to describe this object. Many methods, of supervised classification, are approached including Naïve Bayesian networks [16], decision trees [17] and support vector machines (SVM) [18, 19]. Researches demonstrate the efficiency of the Naïve Bayesian networks to accomplish a relevant classification. However, they are faced with problems when they cope with imperfection, as they are based on probability theory. We briefly present in the following an overview of some probabilistic classifiers.

2.1   Naïve Bayes

A naïve Bayes classifier applies Bayes formula that computes probability measures [16]. This probability includes the frequency of values, in the training set, and combinations of values with independence assumptions. In other terms, the naïve Bayes classifier presumes that the occurrence of a specific class feature is independent of the existence of any other feature. That is, all terms are unrelated to each other given a class. This classifier is trained in a supervised mode. Naïve

Bayesian models mainly compute the probability of a class based on the distribution of the words in the training set [20]. Particular advantages of naïve Bayes classifier are that it involves minor amount of training data to determine the measures required for classification. Naïve Bayes are applied for binary and multiclass classification.

2.2 Decision Trees

Decision trees present a learning method that uses a tree to model the decision potential consequences. They apply a hierarchical division of the basic dataset with the use of diverse text attributes [20]. A tree node describes test on an attribute. A branch defines a test result and a leaf represents a class value (a decision). Decision trees are accomplished in two steps: (i) building (i.e. training) and (ii) pruning (i.e. test or classification). As input, provided data are described by the same attributes of the built tree. This tree is parsed according to the related data. We divide these data using breadth-first or depth-first approaches until we assign a particular class for the input data.

2.3 Support Vector Machines

Support Vector Machines (SVMs) are presented by Vapnik [19] for machine learning. They are used to analyze morphological features of natural languages [3, 26]. They are included in Arabic text classification by Mesleh [21]. SVM classifiers divide the data space among the classes using linear or non-linear delimitations [20]. SVM classifier starts from a set of training instances, each one is assigned to a class category. A training model is built to assign one class to a new given instance. In fact, an instance is represented as a point in space. SVM supports high dimensional spaces. SVM method is flexible and can be joined with interactive methods [15].

3 Possibilistic Classifier

The Naïve Possibilistic Network Classifier was proposed by Haouari et al. [8] to deal with incomplete information [12, 13] of training and test datasets (i.e. uncertain and imprecise). This classifier is based on the possibility theory [10, 11]. Therefore, we resume in the following the major measures used for the classification procedure. In the classification task, the first step defines the training phase. It consists in identifying rules from the available knowledge. We build throughout instances described by values of training attributes. The second step is the test phase. We classify the non-classified instances according to the rules already obtained.

3.1 The Training Phase

In the training set, data are presented by instances whose class values are identified. The training set is imperfect whenever the attributes or classes give uncertain and/or imprecise information. An instance is imprecise when it gives more than one value for an attribute. It is uncertain when the class has more than one value. This imperfection is supported by the Naïve Possibilistic Classifier. Haouari et al. [8] compute possibility means (called π_{Pm}) to denote the imperfection. To each attribute a_j, describing the training set, we determine its possibility means $\pi_{Pm}(a_j|c_i)$ [8] related to the class value c_i according to the formula:

$$\pi_{Pm}(a_j|c_i) = \underset{T(a_j,c_i)}{\text{mean}}\ \beta_j * \pi(I_k|c_i) \tag{1}$$

The $\underset{T(a_j,c_i)}{mean}$ identifies the average of the associated expression computed over a training set named T. The measure $\beta_j = 1/n$ denotes the imprecision rate of the attribute a_j. The n determines the cardinal of attribute values (i.e. a_j) [8]. $\pi(I_k|c_i)$ is the possibility distribution of the attribute values I_k of the instance k given the class c_i [8].

We may realize that particular values of a given attribute ensure better impact in defining the correct class. Possibility theory represents this fact through the necessity measure [25]. The necessity measure N is derived from the possibility measure Π:

$$N(A) = \underset{\omega \notin A}{\min}[1 - \pi(\omega)] = 1 - \Pi(\overline{A}) \tag{2}$$

Where $\Pi(A)$ presents the possibility measure and is equal to:

$$\Pi(A) = \underset{\omega \in A}{\max} \pi(\omega) \tag{3}$$

A describes the subset of states counted in the universe of discourse named Ω. π is the possibility distribution. The $\overline{A}$ denotes the complement of A, i.e. the elements of Ω that do not belong to the event A. We prepare, for each morphological feature, a list of possibility and necessity measures associated to its training set.

3.2   The Test Phase

To classify an imperfect testing set, we compute the possibility distribution of each class c_i given the set of attribute values of the imperfect instance I_k having m attributes [6, 7]:

$$\Pi(c_i|I_k) = \prod_{j=1}^{m} \beta_j * \pi_{Pm}(a_j|c_i) \tag{4}$$

The $\pi_{Pm}(a_j|c_i)$ is calculated through the initial training. If the instance is perfect, β_j (the number of the attribute values) is equal to 1. We compute the necessity measure of the class c_i according to the instance I_k. A given instance is assigned to the most plausible class $c*$. Ayed et al. in [6] use, only, the possibility distributions to determine $c*$. We propose, in this work, to include the sum of necessity and possibility distributions as follows:

$$c^* = \underset{c_i}{\arg\max}\ (\Pi(c_i|I_k) + N(c_i|I_k)) \tag{5}$$

4   The Classification Procedure

We search in our approach the best classifier to disambiguate the morphological features of Arabic texts. The ambiguity of Arabic words generates imperfect training and test datasets. Possibilistic classifier deals with such problems and treats imperfect data. However, probabilistic ones incorporate only perfect data. Hence, we propose to transform the imprecise and uncertain dataset in order to make them available for use to the probabilistic classifiers.

4.1   Data Transformation

We start with presenting an imperfect dataset. Table 1 gives an example of a training set. We assume that the class to disambiguate is POS and the attributes are POS-1 and POS+1. The training set, in Table 1, is composed of 4 instances. The first instance is uncertain since it provides two possible class values (NOUN, NOUN-PROP). The second instance presents imprecision because it gives two values of the attribute POS-1 (NOUN, VERB-PERFECT). The possibilistic classifier may handle this dataset as input for the training procedure.

Table 1. Imperfect instances of a training dataset

POS-1	POS+1	POS (class)
NOUN	VERB-PERFECT	{NOUN; NOUN-PROP}
{NOUN;VERB-PERFECT}	NOUN-PROP	NOUN
NOUN-PROP	NOUN	NOUN
NOUN	NOUN	VERB-PERFECT

We transform the data structure in order to obtain a perfect dataset without loosing sense. To resolve the imprecision problem, we designate the values, of the attribute A, by $A_i = \{a_1, a_2, ..., a_n\}$. From this set, we produce new attributes' denotations. Indeed, we associate the attribute A with each value a_i to form new attribute called "A_a_i".

So, the attribute POS-1 has 3 possible values (NOUN, VERB-PERFECT and NOUN-PROP) in the previous dataset (see Table 1). We get, then, 3 attributes (POS-1_NOUN, POS-1_VERB-PERFECT and POS-1_NOUN-PROP). We give, to the new attributes, binary values (0 or 1). For a given instance, if a_i belongs to the values of the attribute A then the attribute "A_a_i" is equal to 1. From the dataset given in Table 1, we build a new dataset given in Table 2.

Table 2. A dataset with precise instances

POS-1_ NOUN	POS-1_ VERB-PERFECT	POS-1_ NOUN-PROP	POS+1_ NOUN	POS+1_ VERB-PERFECT	POS+1_ NOUN -PROP	POS (class)
1	0	0	0	1	0	{NOUN; NOUN-PROP}
1	1	0	0	0	1	NOUN
0	0	1	1	0	0	NOUN
1	0	0	0	1	0	VERB-PERFECT

To resolve the class uncertainty, we propose to decompose one instance into others having one class value. If an instance has n possible class values $\{c_1, c_2, \ldots, c_n\}$, then we get n associated instances each one has the same attribute values and the class c_i. The certain instances (having one class value) will be duplicated in order to increase their weights in computing the classification measures. Table 3 presents a perfect dataset generated from those given in Table 2.

Table 3. A precise and certain dataset

POS-1_ NOUN	POS-1_ VERB- PERFECT	POS-1_ NOUN- PROP	POS+1_ NOUN	POS+1_ VERB- PERFECT	POS+1_ NOUN -PROP	POS (class)
1	0	0	0	1	0	NOUN
1	0	0	0	1	0	NOUN-PROP
1	1	0	0	0	1	NOUN
1	1	0	0	0	1	NOUN
0	0	1	1	0	0	NOUN
0	0	1	1	0	0	NOUN
1	0	0	0	1	0	VERB-PERFECT
1	0	0	0	1	0	VERB-PERFECT

Therefore, we acquire perfect data that can be classified using any probabilistic classifier.

4.2   Framework Architecture

We present the general proposed framework to disambiguate non-vocalized Arabic texts. This framework provides training sets from vocalized texts. We illustrate, in Figure 1, the process of classification tasks for the possibilistic and probabilistic classifiers. We use Arabic vocalized texts as input data. For each word of the text, we compute the values of 14 morphological features (MF) as in [6, 7]. Therefore, we use an extended version of the morphological analyzer *AraMorph* [6, 7] that handles vocalized and non-vocalized texts.

Preparing data consists in reorganizing the ambiguous analyzed data to obtain instances for test and training. Each instance corresponds to a word w. For each morphological feature F, we produce a dataset. The attributes are $MF-2$, $MF-1$, $MF+1$ and $MF+2$. They respectively define the morphological features of the two preceding and following terms of the word w in a given sentence. For example DET-2 and DET-1 are the attributes describing the DETERMINER values of the two previous words. To describe the morphological feature' class F, we include 56 attributes (14x4).

We notice that the imprecision appears more on the test set as we use non-vocalized texts. However, even vocalized words might provide ambiguous instances. For example, the word "أَحْمَدُ" provides two values of the POS (part-of-speech) feature i.e. verb (I thank) and proper noun (Ahmad).

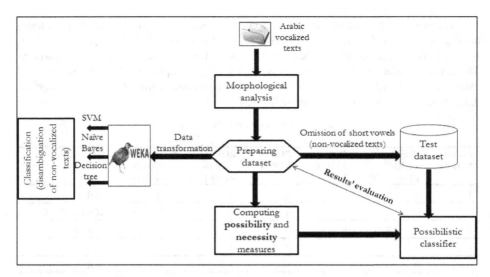

Fig. 1. The architecture of the classification procedure

To classify Arabic texts, we omit their short vowels to get non-vocalized texts. These texts are extracted from the dataset already defined. For each word, we produce its instance described by the same attributes used for the training set. We obtain for each non-vocalized word of a text 14 instances corresponding to all the morphological features. Disambiguation involves associating the appropriate value of each feature. To classify texts using possibilistic classifier, we compute possibility and necessity measures over the training sets [6, 7]. Referring to the measures previously calculated, we determine the best class value for each morphological feature (cf. formula 5). We evaluate disambiguation rates by comparing the disambiguation results of non-vocalized words to the original vocalized version.

To disambiguate texts using the SVM, Naïve Bayes and Decision Tree classifiers [27], we transform imperfect data (cf. Section 4.1) into compatible input format for the machine learning software WEKA[1]. WEKA provides the number of correctly classified instances given by SVM, Naïve Bayes and Decision Tree.

5   Experimental Evaluation

We achieve our experiments on the vocalized Arabic corpus of Hadiths' texts. This corpus was used in various research works [18] and many studies attempt to classify Hadith's texts [23, 24]. The corpus is composed of six books: Sahih Al-Bukhari, Sahih Muslim, Sunan Ettermidhi, Sunan Ibn Majah, Sunan Annasaii, and Sunan Abi Dawud [22]. It presents a big data collection composed of more than 2.5 million words. This corpus covers about 90000 data among titles and paragraphs.

---

[1] http://weka.wikispaces.com/

We generate 14 training sets each one is related to a morphological feature (POS, determiner, gender, voice, aspect, etc.). We calculate the sum of possibility and necessity measures over the built training sets. We classify the non-vocalized texts based on the imperfect instances given by Arabic Hadith texts (cf. Section 3). Thus, we use the cross-validation method to assess the possibilistic and probabilistic classifiers. Indeed, we form 10 iterations: 90% of a vocalized text is used for training and we eliminate short vowels of 10% to be classified.

Table 4 illustrates disambiguation rates of the morphological features given by SVM, Decision Trees and Naïve Bayesian 4 classifiers.

Table 4. Disambiguation rates of 14 morphological features

Morphological Feature	Decision Tree classifier	SVM classifier	Naïve Bayesian classifier	Possibilistic classifier
POS	89.58 %	89.98 %	88.62 %	90.34 %
ADJECTIVE	96.51 %	96.51 %	96.51 %	97.63%
ASPECT	71.20%	71.20%	71.20%	79.19%
CASE	56.12%	56.12 %	56.12 %	63.36%
CONJUNCTION	83.03 %	83.03 %	83.03 %	82.74%
DETERMINER	64.16 %	64.12 %	64.12%	95.33%
GENDER	57.15%	57.15 %	57.15 %	93.66%
MODE	99.32 %	99.32 %	99.32 %	99.96%
NUMBER	85.18 %	85.18 %	85.18 %	90.91%
PARTICLE	96.65%	96.65 %	96.65 %	98.87%
PERSON	60.22 %	60.22 %	60.22 %	65.22%
PREPOSITION	82.87%	82.87 %	82.87 %	85.80%
VOICE	71.21 %	71.21 %	71.21 %	79.05%
PRONOUN	55.84 %	56.88 %	55.02 %	58.79%
Average	**76.36 %**	**76.46 %**	**76.23 %**	**84.34 %**

Experiments prove that possibilistic classifier gives better disambiguation rates compared with the SVM, Naïve Bayes and Decision Tree classifiers. It presents an average of 84.34% of accurate classified non-vocalized instances. Some morphological features give same results for the probabilistic classifiers. This may be explained by the fact that the associated morphological features give few numbers of class values (not exceeding 6 each). On the other side, the feature PRONOUN (for example) offers about 64 class values which may generate various results from the different classifiers.

6   Conclusion

We presented in this paper a comparative study between a possibilistic classifier and some existing probabilistic classifiers that aimed to disambiguate Arabic morphological features of non-vocalized texts. We described the framework architecture used to train and test texts by defining an approach based on possibility theory that deals with imprecise and uncertain data. Comparing results of the different classifiers, we concluded that possibilistic theory gave better disambiguation rates of the morphological features. Thus, possibilistic theory presented an enhanced way to solve the Arabic morphology problem.

We propose, as future work, to distinguish to which degree a given morphological feature is involved in the disambiguation of another. We suggest adding some statistical measures assessing the impact of each attribute in the classification procedure.

References

1. Khoja, S.: APT: Arabic part-of-speech tagger. In: Proceedings of Student Workshop at the Second Meeting of the North American Association for Computational Linguistics. Carnegie Mellon University, Pennsylvania (2001)
2. Hajic, J.: Morphological Tagging: Data vs. Dictionaries. In: Proceedings of the 1st North American Chapter of the Association for Computational Linguistics Conference, pp. 94–101. Association for Computational Linguistics, Stroudsburg (2000)
3. Roth, R., Rambow, O., Habash, N., Diab, M., Rudin, C.: Arabic Morphological Tagging, Diacritization, and Lemmatization Using Lexeme Models and Feature Ranking. In: Proceedings of the Association for Computational Linguistics Conference (ACL), Columbus, Ohio, USA, pp. 117–120 (2008)
4. Habash, N., Rambow, O.: Arabic Tokenization, Part-of-speech Tagging and Morphological Disambiguation in One Fell Swoop. In: Proceedings of the 43rd Annual Meeting on Association for Computational Linguistics, Stroudsburg, PA, USA, pp. 573–580 (2005)
5. Habash, N., Rambow, O.: Arabic Diacritization Through Full Morphological Tagging. In: Human Language Technologies: The Conference of the North American Chapter of the Association for Computational Linguistics, Stroudsburg, PA, USA, pp. 53–56 (2007)
6. Ayed, R., Bounhas, I., Elayeb, B., Evrard, F.: Bellamine Ben Saoud, N.: A Possibilistic Approach for the Automatic Morphological Disambiguation of Arabic Texts. In: Hochin, T., Lee, R. (eds.) Proceedings of the 13th ACIS International Conference on Software Engineering, Artificial Intelligence, Networking and Parallel Distributed Computing (SNPD), Kyoto, Japan, pp. 187–194 (2012)
7. Ayed, R., Bounhas, I., Elayeb, B., Evrard, F., Saoud, N.B.B.: Arabic Morphological Analysis and Disambiguation Using a Possibilistic Classifier. In: Huang, D.-S., Ma, J., Jo, K.-H., Gromiha, M.M. (eds.) ICIC 2012. LNCS (LNAI), vol. 7390, pp. 274–279. Springer, Heidelberg (2012)
8. Haouari, B., Ben Amor, N., Elouedi, Z., Mellouli, K.: Naïve Possibilistic Network Classifiers. Fuzzy Sets and Systems 160(22), 3224–3238 (2009)
9. Dubois, D., Prade, H.: Possibility Theory: An Approach to computerized Processing of Uncertainty. Plenum Press, New York (1994)
10. Dubois, D.J., Prade, H.: Théorie des possibilités: applications à la représentation des connaissances en informatique. Masson, Paris (1985)
11. Dubois, D., Prade, H.: Possibility Theory: Qualitative and Quantitative Aspects. In: Gabbay, D.M., Smets, P. (eds.) Handbook on Defeasible Reasoning and Uncertainty Management Systems, pp. 169–226. Kluwer Academic, Springer, Dordrecht, Netherlands (1998)

12. Alkuhlani, S., Habash, N., Roth, R.: Automatic Morphological Enrichment of a Morphologically Underspecified Treebank. In: Clemmer, A., Post, M. (eds.) Proceedings of the Conference of the North American Chapter of the Association for Computational Linguistics: Human Language Technologies, HLT-NAACL, pp. 460–470. Omnipress of Madison, Wisconsin (2013)
13. Bounhas, M., Mellouli, K., Prade, H., Serrurier, M.: Possibilistic classifiers for numerical data. Soft Computing 17(5), 733–751 (2013)
14. Buckwalter, T.: BuckwalterArabicMorphological Analyzer Version 2.0. Linguistic Data Consortium (LDC) catalogue number LDC2004L02 (2004) ISBN 1-58563-324-0
15. Raghavan, H., Allan, J.: An interactive algorithm for asking and incorporating feature feedback into support vector machines. In: ACM SIGIR Conference (2007)
16. Pearl, J.: Probabilistic reasoning in intelligent systems: networks of plausible inference. In: Probabilistic Reasoning in Intelligent Systems. Morgan Kaufmman, San Francisco (1988)
17. Quinlan, R.: Induction of decision trees. Machine Learning 1, 81–106 (1986)
18. Harrag, F., Hamdi-Cherif, A., Malik, A., Al-Salman, S., El-Qawasmeh, E.: Experiments in Improvement of Arabic Information Retrieval. In: Proc. 3rd International Conference on Arabic Language Processing (CITALA), Rabat, Morocco, pp. 71–81 (2009)
19. Vapnik, V.: Statistical Learning Theory, pp. 1–736. Wiley, New York (1998)
20. Aggarwal, C.C., Changing, Z.: A survey of text classification algorithms. In: Mining Text Data, pp. 163–213 (2012)
21. Mesleh, A.: Support Vector Machines based Arabic Language Text Classification System: Feature Selection Comparative Study. In: 12th WSEAS International Conference on applied mathematics, Cairo, Egypt, pp. 11–16 (2007)
22. Al-Echikh, A.A.: Encyclopedia of the six major citation collections. Daresselem, Ryadh (1998)
23. Jbara, K.: Knowledge Discovery in Al-Hadith Using Text Classification Algorithm. Journal of American Science 6(11), 409–419 (2010)
24. Alkhatib, M.: Classification of Al-Hadith Al-Shareef Using Data Mining Algorithm. In: Proceedings of European, Mediterranean & Middle Eastern Conference on Information Systems, Abu Dhabi (2010)
25. Bounhas, I., Elayeb, B., Evrard, F., Slimani, Y.: Organizing Contextual Knowledge for Arabic Text Disambiguation and Terminology Extraction. Knowledge Organization 38(6), 473–490 (2011)
26. Outahajala, M., Benajiba, Y., Rosso, P., Zenkouar, L.: POS Tagging in Amazighe Using Support Vector Machines and Conditional Random Fields. In: Muñoz, R., Montoyo, A., Métais, E. (eds.) NLDB 2011. LNCS, vol. 6716, pp. 238–241. Springer, Heidelberg (2011)
27. Georgescul, M., Rayner, M., Bouillon, P.: Spoken Language Understanding via Supervised Learning and Linguistically Motivated Features. In: Hopfe, C.J., Rezgui, Y., Métais, E., Preece, A., Li, H. (eds.) NLDB 2010. LNCS, vol. 6177, pp. 117–128. Springer, Heidelberg (2010)

Forecasting Euro/Dollar Rate with Forex News[*]

Olexiy Koshulko[1], Mikhail Alexandrov[2,3], and Vera Danilova[2,3]

[1] Glushkov Institute of Cybernetics, NASU, Ukraine
koshulko@gmail.com
[2] Autonomous University of Barcelona, Spain
[3] Russian Presidential Academy of National Economy and Public Administration, Russia
MAlexandrov@mail.ru, maolve@gmail.com

Abstract. In the paper we build classifiers of texts reflecting opinions of currency market analysts about euro/dollar rate. The classifiers use various combinations of classes: growth, fall, constancy, not-growth, not-fall. The process includes term selection based on criterion of word specificity and model selection using technique of inductive modeling. We shortly describe our tools for these procedures. In the experiments we evaluate quality of classifiers and their sensibility to term list. The results proved to be positive and therefore the proposed approach can be a useful addition to the existing quantitative methods. The work has a practical orientation.

Keywords: GMDH, text classification, Forex.

1   Introduction

Players of Forex market use various mathematical and heuristic methods for forecasting currency rate. One can find their general descriptions on many Web pages. But the "devil is hidden in details" and skilled Forex players keep these details a secret. Meanwhile there is an additional free of charge source of information in Internet, which could be useful for forecasting.

In the paper, we consider the possibility to forecast a behavior of the currency pair euro/dollar using texts of Forex news. Forex news taken together during 2 days are considered as one textual unit. Usually such a unit includes 3-5 documents. The relative frequencies of terms (keywords) from these documents are independent variables for a model to be built. The tendency of changes in currency rate is a subject of forecast, so these changes are considered as a dependent variable. We consider 4 types of classifiers presented in the Table 1 and build a model for each of them.

To select terms we use the procedure based on criterion of word specificity. Such a criterion takes into account the relative frequency of word occurrence in a given corpus and in some standard corpus. To build the model itself we use the Group Method of Data Handling (GMDH). This method builds the model of optimum complexity from a given class using training and control set of an initial data. In the paper we shortly

---

[*] Work done under partial support of the British Petroleum grant (RPANEPA-S1/2013).

E. Métais, M. Roche, and M. Teisseire (Eds.): NLDB 2014, LNCS 8455, pp. 148–153, 2014.
© Springer International Publishing Switzerland 2014

Table 1. Types of classifiers

Type	Classes
1	Growth, Fall
2	Growth, Constancy, Fall
3	Growth, Not-growth
4	Fall, Not-fall

describe these methods and our tools. Forex news are taken from the well-known and a reliable source. We dealt with 89 textual units covering almost 9 months 2012-2013.

In the paper [3] we described our first experience with using text classifiers for forecasting currency rates. Unlike [3] in this paper we focus on the linguistic resources. The criterion of word specificity is a well known among NLP specialists. The behavior of this criterion and free share tool is presented in detail in [7]. GMDH has already demonstrated its effectiveness in several NLP applications. We could mention here: testing word similarity and building word frequency list [1], subjectivity/sentiment analysis [2], classification of medical documents [5], etc.

The other sections of the paper are organized as follows. Section 2 describes linguistic resources (corpus and vocabularies). Section 3 describes the technique of inductive modeling. Short section 4 presents the results of experiments. Section 5 concludes the paper.

2   Linguistic Variables

2.1   Corpus of News

The Forex news were downloaded from the site http://www.dailyfx.com/. This site proved to be the most confident source of information. The example of news is presented in the Table 2.

Table 2. Example of Forex news

Measures changes in sales of the German retail sector. Given that consumption makes up a significant portion of German GDP, the Retail Sales figure can act as an indicator of domestic demand. High or rising Retail Sales may spur German consumption, translating into economic growth. However, uncontrolled growth runs the risk of inflationary pressures. Since Germany is a large part of the Euro-zone, German figures may have some impact on the market. The headline figure is expressed in percentage change in the value of sales.

As we have mentioned above each textual unit contains news taken during 2 days. The general characteristics are presented in the Table 3.

To construct a learning document set we fix a rate just after these days. Table 4 shows example of data. Speaking constancy we mean an insignificant change in currency rate. The threshold for such an evaluation depends on expert opinion. In our classifiers we use also combined categories Not-growth and Not-fall. Obviously that Not-growth = {Constancy, Fall} and Not-fall = {Constancy, Growth}.

Table 3. General characteristics of the corpus

Characteristics	Value
Number of documents (textual units)	89
Average lengths of document (words)	513

Table 4. Source data for learning model (example)

Date	Documents	Euro/Dollar rate	Market
13.03.2013	Text-1	1,25	Growth
15.03.2013	Text-2	1,15	Fall
17.03.2013	Text-3	1,24	Growth
19.03.2013	Text-4	1,22	Constancy
21.03.2013	Text-5	1,26	Growth

The structure of document corpus for different classes is presented in Table 5. One should note that for the binary classifier {Growth, Fall} the texts from the class Constancy were distributed between the classes Growth and Fall.

Table 5. Structure of document corpus

Type	Growth	Constancy	Fall	Not growth	Not fall	Totally
Type 1	43	-	46	-	-	89
Type 2	30	24	35	-	-	89
Type 3	30	-	-	59	-	89
Type 4	-	-	35	-	54	89

2.2 Lists of Terms

To transform the documents to their numerical form we select terms. In this paper, we use one-word terms. The procedure contains 2 steps:

Step 1. Automatic word selection with the program *LexisTerm;*

Step 2. Expert corrects the selected words removing redundant words and adding the necessary ones.

LexisTerm selects words according the criterion of word specificity. Speaking 'word specificity' with respect to a given corpus we mean a factor $K \geq 1$, which shows how much word frequency in the corpus $f_C(w)$ exceeds its frequency in any standard corpus $f_L(w)$: $K = f_C(w) / f_L(w)$. In our work we use General Lexis of English that reflects word frequencies in the British National Corpus. This lexis is available in Internet.

Program *LexisTerm* is a tool developed with the participation of one of the authors. It is described in detail in [7] and it is widely used in our projects in Spain, Peru and Russia.

We compared contents of lists, which were built by the program for K=2,5,10,20, 50 and founded that: when K<10 the lists included many insignificant terms, when

$K>10$ we lost many useful terms. So, we took the threshold $K=10$ as the basic value. Speaking about the series of K-values we should say that only with these 'logarithmic' steps we could see the essential changes in the contents of lists.

With $K=10$ we obtained 155 terms. The most frequent terms from the list are: *inflation, german, figure, economic, account, consume, trade, changes, goods, growth, rates, prices,* etc. This list was then corrected by an expert, and the resultant list contained 86 terms. These terms can be named linguistic variables.

In order to study the sensibility of results to size and contents of term list we repeated term selection with $K=50$. Using this threshold we obtained 18 terms and after correction the final list included 14 terms.

The results of these experiments are presented in the Table 6. The last column contains the lists of terms in the form of stems.

Table 6. Term selection (number of terms)

Threshold	LexisTerm (terms)	Final list (terms)	Examples of terms
$K = 10$	155	86	account addition appreciate balance boost breakdown capital chang confiden consumer consumption contribute control currency…
$K = 50$	18	14	confiden consumer conversely decline deficit export import increase index indicat inflation manufacture negative pressur

The selected terms allowed us to parameterize texts for our experiments. All vectors are normalized on one. So, document representation doesn't depend on document size.

3   Inductive Modeling

3.1   Group Method of Data Handling

To build classifiers we used technique of inductive modeling presented by the Group Method of Data Handling (GMDH). The founder of GMDH is the remarkable Ukrainian scientist O. Ivakhnenko. His first International publication devoted to GMDH as long ago as 1971 [4]. The description of GMDH and its applications are presented in [9,10]. Here is the simplest realization of GMDH:

(1) An expert defines a sequence of models, from the simplest to more complex ones.
(2) Experimental data are divided into two data sets: training data and control data
(3) For a given kind of model, the best parameters are determined with training data.
(4) This model is tested on controñ data using any external criteria
(5) The external criteria (or the most important one) are checked on having reached a stable optimum. In this case the search is finished. Otherwise, more complex model is considered and the process is repeated from the step 3.

The typical form of model presentation is the polynomial one:

$$y = a_0 + \Sigma b_{ij} x_i x_j + \Sigma c_{ijk} x_i x_j x_k +$$

Here: y is a dependent variable, x_i are independent variables, a,b,c are coefficients to be determined. We can use both positive and negative power functions as $x^m{}_i$, $m<0$

3.2   GMDH Shell

GMDH Shell (GS) is a well-known tool for time series prognosis, function approximation and object classification including extended possibilities for visualization of results [8]. One of the authors is the main developer of this software. GS employs a technique of GMDH. At present GS includes two algorithms:

- Combinatorial GMDH;
- GMDH-type neural networks.

In our research we use the classification option. GS includes here the well-known One-vs-All method [11]. This method reduces multiclass classification to binary classification. Each binary classifier is presented in the form of an equation with linguistic variables we discussed above. Inductive modeling allows to find an equation of optimal complexity.

The value of the equation determines the level of confidence to a given class. The class having the largest level of the confidence is taken as a winner. Obviously in case of two classes only one classifier is needed.

4   Experiments

We completed experiments with two data sets related to long term list and short term list respectively. To adjust the GS we used recommendations from [6]. Here is characteristics of GS: combinatorial algorithm, model complexity is less or equal 3, members include only w_i, $w_i w_j$, $w^2{}_i$, where w_i, w_j are terms from the lists (Table 6). As the example, we show the model for classifier {Growth, Fall} with the short list:

$$y = 0.57 + 2.58\ w_3 w_{12} - 3.19\ w_5 w_7 - 2.76\ w_8 w_{13}$$

The rule for decision-making is: class Growth if $y \geq 0$ and class Fall if $y < 0$.

Table 7 presents the results of experiments for each of classifiers. These results refer to accuracy on control data set (but not on all data set).

Table 7. Results of experiments for all classifiers

Type	Long list Accuracy %	Short list Accuracy %	Baseline %
Growth, Fall	79	71	52
Growth, Constancy, Fall	68	57	39
Growth, Not-growth	72	67	66
Fall, Not-fall	68	64	60

5   Conclusions

Results. We proposed a new way for forecasting rate of currency pair based on the analysis of news of Forex market. The preliminary results proved to be positive for simple (not combined) classes both for large and short term lists. The proposed way can't substitute the existing strategies on currency market but it can be a useful addition to them. Speaking 'addition' we mean support of various hypotheses about currency rate.

Future Work. We intend to implement experiments with essentially larger data set of euro/dollar rate. We also intend to repeat experiments with other currencies.

Acknowledgment. The authors thank B.Sc. Timur Garaev for his help in building the term vocabularies that allowed to improve essentially the quality of results.

References

1. Alexandrov, M., Blanco, X., Makagonov, P.: Testing Word Similarity: Language Independent Approach with Examples from Romance. In: Meziane, F., Métais, E. (eds.) NLDB 2004. LNCS, vol. 3136, pp. 229–241. Springer, Heidelberg (2004)
2. Alexandrov, M., et al.: Inductive Modeling in Subjectivity/Sentiment Analysis (case study: dialog processing). In: Proc. of 3rd Intern. Workshop on Inductive Modeling (IWIM 2009), Krynica, Poland, pp. 40–43 (2009)
3. Garaev, T., Alexandrov, M., Koshulko, O.: Text classifier as a tool for short-term forecast of currency rates. In: Proc. of 4th Intern. Conf. on Inductive Modeling (ICIM 2013), pp. 261–266. NAS of Ukraine, Prague Tech. University, Kyev (2013)
4. Ivakhnenko, A.: Polynomial theory of complex systems. IEEE Transactions on Systems, Man, and Cybernetics SMC-1(4), 364–378 (1971)
5. Kaurova, O., Alexandrov, M., Koshulko, O.: Constructing classifiers of medical records presented in free text form. In: Proc. of 4th Intern. Conf. on Inductive Modeling (ICIM 2013), pp. 273–278. NAS of Ukraine, Prague Tech. University, Kyev (2013)
6. Koshulko, O., Koshulko, G.: Validation Strategy Selection in Combinatorial and Multilayered Iterative GMDH Algorithms. In: Proc. Intern. Workshop on Inductive Modeling (IWIM 2011), pp. 51–54. NAS of Ukraine, Prague Tech. University, Kyev (2011)
7. Lopez, R., et al.: Lexisterm – the program for term selection by the criterion of specificity. In: Artificial Intelligence Applications to Business and Engineering Domain, vol. 24, pp. 8–15. ITHEA Publ., Rzeszov (2011)
8. Program GMDH Shell, http://gmdhshell.com
9. Madala, H., Ivakhnenko, A.: Inductive learning algorithms for complex systems modelling. CRC Press (1994)
10. Stepashko, V.: Ideas of academician O. Ivakhnenko in Inductive Modeling field from historical perspective. In: Proc. of 4th Intern. Conf. on Inductive Modeling (ICIM 2013), pp. 31–37. NAS of Ukraine, Prague Tech. University, Kyev (2013)
11. Wikipedia, One-vs-All,
    http://en.wikipedia.org/wiki/Multiclass_classification

Towards the Improvement of Topic Priority Assignment Using Various Topic Detection Methods for E-reputation Monitoring on Twitter

Jean-Valère Cossu, Benjamin Bigot, Ludovic Bonnefoy, and Grégory Senay

LIA/Université d'Avignon et des Pays de Vaucluse
39 chemin des Meinajaries, Agroparc BP 91228, 84911 Avignon cedex 9, France
firstname.name@univ-avignon.fr
http://lia.univ-avignon.fr/

Abstract. Topic priority assignment is defined in *RepLab-2013* as labelling a topic according to its level of priority (ALERT, MILDLY IMPORTANT or UNIMPORTANT) in order to highlight topics requiring immediate attention for online reputation monitoring. Although they are strongly linked, topic detection and priority assignment have been previously treated as separate tasks. We study the impact of integrating topic detection outputs in the process of topic priority assignment.

1 Introduction

The amount and richness of the information collectively generated by users on online social networks have increased drastically during these last years. It is now well established that online social interactions often reflect in real-time the impact of real-world events on people opinions. Understanding social events is therefore crucial for persons and companies concerned with their online reputation. Companies typically spend a lot of money to get reliable satisfaction polls using call centers and surveys, and online social networks are certainly carrying key information to anticipate and react to the versatility of public opinions. Considering this amount of documents, automatic approaches are needed and have to deal with many sources of noise and perturbations. Noisy data mainly results from entity names ambiguities (e.g. jaguar: animal/car manufacturer), and an important number of linguistic variants and para-linguistic phenomena.

Replab 2013[1] provides a framework to evaluate Online Reputation Management systems on Twitter. The organizers have decomposed the monitoring issue into 4 subtasks: filtering, polarity classification, topic detection and priority assignment. In this paper, we are interested in 2 tasks: Topic Detection in which systems have to group together tweets related to one entity (a person, a company, etc.) by subject/event/conversation; and Priority Assignment consisting in ranking topics by priority (ALERT, MILDLY IMPORTANT and UNIMPORTANT). We will investigate the combination of these 2 subtasks in order to improve the quality of priority assignment for reputation monitoring.

---

[1] http://www.limosine-project.eu/events/replab2013

E. Métais, M. Roche, and M. Teisseire (Eds.): NLDB 2014, LNCS 8455, pp. 154–159, 2014.
© Springer International Publishing Switzerland 2014

2   Related Work

Previous works on topic detection and characterization in tweet collections and streams aim at extracting messages requiring a attention from a user for instance by extracting new events [1], performing trend detection [2] or detecting late-breaking news [3] over the Twitter stream. To our knowledge, most of the contributions to reputation monitoring on Twitter have been proposed in the context 2012 and 2013 editions of Replab, with methods based on unsupervised clustering algorithms and supervised classification methods. Similarity between tweet content after a preprocessing consisting in a concept term expansion of filtered tweets words is used in [4]. Three topic detection approaches have been proposed in [5] and [6]: agglomerative clustering using term co-occurrences; agglomerative clustering using a wikified representation of tweet; and a Twitter-Latent Dirichlet Allocationused to discover latent topics in tweets. In [7], both supervised (Naive Bayes and Sequential Minimal Optimization Support Vector Machines) and unsupervised algorithms (K-star) combined with terms selection strategies are used. In [8], Social Network Analysis for tweets clustering is introduced.

Topic priority assignment for reputation monitoring in tweets is similar to topic characterization in Twitter [9]. Most of the contributions have been proposed in the context of Replab and mostly rely on supervised classification methods. In [5], authors use a tweet-level sentiment analysis classifier and exploit the link between priority and polarity values. In [7], three classifiers have been trained using features extracted from tweets content and meta-data.

3   Topic Detection and Priority Assignment Systems

To study the dependencies between the topic detection step and priority assignment, we first propose several systems based either on supervised classification methods or unsupervised clustering algorithms. We also use the **Replab2013 baseline** that consists in tagging the tweets of the test set with the label of the closest tweet (Jaccard word similarity) in the reference.

3.1   Topic Detection Systems

The first method is a **K-means clustering using Jaccard similarity** [10] computed on the overall dataset (training and test tweets). The initial value of K is set to the number of clusters in the training set. As a preprocessing step we remove words appearing only once. The second method is a **Hierarchical clustering using Jaccard similarity** after the same preprocessing. The tree is cut according to the number of clusters in the training set. Our third system is based on a **Maximum *a posteriori* feature selection (MAP)**. This supervised method is based on [11]. Features are words, bigrams, distant bigrams (one gap) and tweet authors. It consists in selecting the most discriminant features for each topic using posterior probabilities of each term for a topic over the training dataset. Topic attribution is done by considering the maximum contribution of a tweet to a topic.

3.2   Priority Assignment Systems

This first approach called the **KBA 2012 system** [12] has been proposed for the Knowledge Base Acceleration (KBA) task in TREC 2012 which is similar to RepLab priority assignment. The main difference lies in the kind of documents processed (web pages versus tweets). This method captures intrinsic characteristics of highly relevant documents using three types of features (document centric features, entitys profile features, and time features). We use two Random Forest classifiers (unimportant versus mildly important and important, them mildly important versus important). It matches a tweet in the test set with the k most similar tweets of the training set. Similarity is computed with Jaccard similarity on discriminant bag-of-words computed on tweet content and metadata (author, entity). k (equal to 6) has been fixed by cross-validation on the training set.

4   Relational Model, Corpus and Metrics

The corpus is a bilingual collection of tweets related to 61 entities from 4 domains: *Automotive, Banking, Universities* and *Music/Artists*. The tweets are labelled with 8 attributes: **tweet_id, author, entity, tweet_content, language, date, category** and **retweet**. The outputs are binary relations among tweet ids:

- **filter** ⊆ **tweet_id** × **entity** ∪ {NULL} ,
- **opinion** ⊆ **tweet_id** × {POSIVE, NEUTRAL, NEGATIVE},
- **priority** ⊆ **tweet_id** × {NONE, MIDLY, ALERT},
- **topic** ⊆ **tweet_id** × **tweet_id** is used to cluster the tweets by similarity.

The only defined functional dependency are **topic** → **entity** → **category** and **topic** → **filter** but **topic, opinion** → **priority** can also be assumed over more than 90% of records. The training set contains 34,496 tweets and the test set 70,412. Clearly, finding the appropriate **topic** relation is not a classification task but a clustering one since the training set contains 3,488 unique topics and the test set 5,343. However, record based NLP machine learning classification approaches appear to be efficient in providing a first approximation and additional attributes that can be further used in clustering.

Among the 3 priority levels, ALERT is the smallest with only 1,540 in the train set (3,161 tweets for test). We can find 17,954 MILDLY IMPORTANT (35,995 in the test) tweets and 31,256 tweets (15,378) are annotated as UNIMPORTANT. Note that most of the ALERT Tweets are related to *Banking*.

Metrics are Accuracy, Reliability (R), Sensitivity (S) and F-measure (based on R&S) [13]. Reliability and Sensitivity can be seen as precision and recall under the assumption that a test dataset can be seen as a bag of relationships (<, >, =) between the priority of test documents. Scoring is achieved by a comparison with the relations held in a gold standard. We have also computed classical F-measure (based on Precision and Recall) for each priority class.

5   Experiments

We consider the output of priority assignment and evaluate its improvement using additional information brought by topic detection. Performances of our topic detection systems, Replab2013's baseline and best system [6] are reported in Table 1. Our methods outperform the baseline and yield different values of Reliability (R) and Sensitivity (S). Two operating points have been set for the MAP (threshold on the number of words for the training) method in order to maximize either Sensitivity (MAP#1) or Reliability (MAP#2). The low performance of clustering methods is caused by significant differences of topics numbers in the both training and test set.

We now compare (cf. Table 2) the performances of priority assignment methods alone, and combined with the topic gold standard. In the first case, priority assignment based on KNN and KBA outperform the baseline and in the second case, adding the topic detection gold standard significantly improves topic priority. KNN now reaches an accuracy equal to 0.69 (+6 points comparing to KNN taken alone) and the best values of F-measure per class. F-measure (based on R&S) is also improved up to 0.387 instead of 0.335. This result proves good topic definitions do contain relevant information that improves priority assignment.

In the next experiment, we combine topic detection and priority assignment methods (cf. Table 3). Beyond the fact that results are lower than priority assignment system taken alone (F-measure=.335 for KNN cf. Table 2), it is

Table 1. Performances of topic detection systems

Method	Reliability	Sensitivity	F-Measure(R&S)
Replab baseline	.152	.217	.173
K-means clustering	.308	.157	.201
Hierarchic clustering	.261	.220	.227
MAP features selection #2	**.381**	.172	.238
MAP features selection #1	.193	**.497**	.266
Best@Replab2013	.462	.324	**.325**

Table 2. Priority assignment alone, with baseline and gold standard topics

Method	F-measure(Prec.&Rec.)			Acc.	Rel.	Sens.	F-m(R&S)
	Alert	Mildly	Unimp.				
Priority assignment only							
Baseline	.336	.643	.617	.530	**.403**	.248	.274
KNN	**.415**	**.684**	.646	**.627**	.387	**.315**	**.335**
KBA	.025	.560	**.705**	.585	.315	.276	.282
Priority assignment + gold standard topic detection							
Baseline	.441	.706	.703	.649	.511	.281	.326
KNN	**.514**	**.733**	.702	**.690**	**.549**	**.345**	**.387**
KBA	.002	.560	**.705**	.612	.532	.269	.329

very interesting to notice that except for the baselines combination, the performances respect **F-m(baseline) < F-m(MAP#1) < F-m(Hierarch.) < F-m(K-means) < F-m(MAP#2)**. F-measures of combined systems are ranked according to the values of topic detection method's Reliability (cf Table 1).

In one last experiment we study the impact of an automatic topic detection on perfect priority assignment by combining our topic-detection methods with the priority gold standard (cf. Tab. 4). We considered the priority gold standard as a system output and tried to propagate the majority priority label to the whole topic cluster. It's interesting to check how much our clusters can degrade

Table 3. Performances of priority assignment combined with topic detection methods

Method	F-measure			Acc.	Rel.	Sens.	F-m(R&S)
	Alert	Mildly	Unimp.				
Priority assignment + baseline topic detection							
Baseline	.336	.643	.617	.530	.403	**.248**	**.274**
KNN	**.376**	**.672**	**.633**	**.550**	.520	.136	.172
KBA	0	.478	**.661**	.489	**.578**	.071	.098
MAP features selection #1							
Baseline	.342	.657	.659	.628	.383	**.151**	**.195**
KNN	**.378**	**.660**	.646	**.632**	.413	.136	.181
KBA	0	.466	**.672**	.568	**.551**	.098	.126
MAP features selection #2							
Baseline	.329	.643	.628	.574	**.406**	.214	.261
KNN	**.373**	**.669**	.636	**.619**	.405	**.249**	**.288**
KBA	.069	.512	**.657**	.561	.361	.171	.217
Hierarchical clustering using Jaccard similarity							
Baseline	**.342**	.642	.631	.584	.378	.174	.214
KNN	.340	**.659**	.631	**.613**	.391	**.195**	**.239**
KBA	.126	.515	**.662**	.567	**.421**	.150	.192
K-means using Jaccard similarity							
Baseline	.338	.635	.625	.570	.392	.206	.253
KNN	**.365**	**.667**	.628	**.612**	**.416**	**.223**	**.269**
KBA	.130	.514	**.661**	.559	.409	.164	.212

Table 4. Impact of topic detection methods using priority assignment gold standard

Method	F-measure			Acc.	Rel.	Sens.	F-m(R&S)
	Alert	Mildly	Unimp.				
Topic detection + Gold standard Priority							
MAP feat. select. #2	.710	**.840**	**.823**	**.812**	**.756**	**.518**	**.602**
Hierarch. clust.	.712	.785	.769	.783	.696	.438	.519
K-means clust.	**.769**	.815	.791	.761	.655	.367	.437
MAP feat. select. #1	.551	.754	.743	.731	.666	.229	.311
Baseline	.535	.763	.727	.634	.657	.198	.262

the priority gold standard. Again we observe that the order of the ranked F-measures is highlighting that Reliability of topic detection seems to have an important effect on the performances of combined systems.

6   Conclusion

We have studied the impact of combining priority classification methods with the outputs of topic detection approaches for the task of topic priority assignment for online reputation monitoring in tweets. Experiments have shown the relevance of this proposition, but actual methods are not yet mature enough to reach better performances than any priority assignment system taken alone. Since such a pipeline approach propagate early stage errors to the later stage we have to study how to tackle this issue with alternative combination strategies or by the use of an unified topic framework which can assign topic and priority in one pass by taking into account both topic and priority predictions.

References

1. Petrovic, S., Osborne, M., Lavrenko, V.: Streaming first story detection with application to Twitter. In: HLT-NACCL, pp. 181–189. ACL (2010)
2. Mathioudakis, M., Koudas, N.: TwitterMonitor: trend detection over the twitter stream. In: SIGMOD 2010, pp. 1155–1158. ACM (2010)
3. Sankaranarayanan, J., Samet, H., Teitler, B., Lieberman, M., Sperling, J.: TwitterStand: news in tweets. In: SIGSPATIAL-GIS, pp. 42–51. ACM (2009)
4. Atif Qureshi, M., O'Riordan, C., Pasi, G.: Concept Term Expansion Approach for Monitoring Reputation of Companies on Twitter. In: CLEF 2012 (2012)
5. Martin-Wanton, T., Spina, D., Amigo, E.: UNED at RepLab 2012: Monitoring Task. In: CLEF 2012 (2012)
6. Spina, D., Carrillo-de-Albornoz, J., Martin, T., Amigo, E., Gonzalo, J., Giner, F.: UNED Online Reputation Monitoring Team at RepLab 2013. In: CLEF 2013 (2013)
7. Sanchez-Sanchez, C., Jimenez-Salazar, H., Luna-Ramirez, W.: UAMCLyR at Replab2013: Monitoring Task. In: CLEF 2013 (2013)
8. Berrocal, J.-L., Figuerola, C., Rodriguez, A.: REINA at RepLab2013 Topic Detection Task. In: CLEF 2013 (2013)
9. Naaman, M., Becker, H., Gravano, L.: Hip and Trendy: Characterizing Emerging Trends on Twitter. Journal of the American Society for Information Science and Technology 62, 5 (2007)
10. Leisch, F.: A toolbox for k-centroids cluster analysis. In: Computational Statistics and Data Analysis (2006)
11. Hazen, T., Richardson, F., Margolis, A.: Topic identification from audio recordings using word and phone recognition lattices. In: ASRU, pp. 659–664. IEEE (2007)
12. Bonnefoy, L., Bouvier, V., Bellot, P.: A Weakly-Supervised Detection of Entity Central Documents in a Stream. In: SIGIR (2013)
13. Amigo, E., Gonzalo, J., Verdejo, F.: A general evaluation measure for document organization tasks. In: SIGIR, pp. 643–652. ACM (2013)

Complex Question Answering: Homogeneous or Heterogeneous, Which Ensemble Is Better?

Yllias Chali[1], Sadid A. Hasan[2], and Mustapha Mojahid[3]

[1] University of Lethbridge, Lethbridge, AB, Canada
`chali@cs.uleth.ca`
[2] Philips Research North America, Briarcliff Manor, NY, USA
`sadid.hasan@philips.com`
[3] IRIT, Toulouse, France
`mustapha.mojahid@irit.fr`

Abstract. This paper applies homogeneous and heterogeneous ensembles to perform the complex question answering task. For the homogeneous ensemble, we employ Support Vector Machines (SVM) as the learning algorithm and use a Cross-Validation Committees (CVC) approach to form several base models. We use SVM, Hidden Markov Models (HMM), Conditional Random Fields (CRF), and Maximum Entropy (MaxEnt) techniques to build different base models for the heterogeneous ensemble. Experimental analyses demonstrate that both ensemble methods outperform conventional systems and heterogeneous ensemble is better.

Keywords: Complex Question Answering, Homogeneous Ensemble, Heterogeneous Ensemble.

1 Introduction

This paper is concerned with the application of ensemble based methods for the complex question answering task. We use query-focused supervised extractive multi-document summarization technique for this purpose [1–3]. Ensemble methods are learning algorithms that construct a set of classifiers and then classify new data points by taking a (weighted) vote of their predictions [4]. Generation of ensembles can be categorized into two types: 1) homogeneous, if the base learning model is built from the same learning algorithm, and 2) heterogeneous, where different learning algorithms are combined to generate the base learning models [12]. Many methods for constructing ensembles have been developed over the years which consider Bayesian voting, manipulation of the training examples, input features and output targets, injecting randomness and so on [2, 6, 14]. The next section presents our experimental design and evaluation framework, and then we conclude the paper with future directions.

E. Métais, M. Roche, and M. Teisseire (Eds.): NLDB 2014, LNCS 8455, pp. 160–163, 2014.

2 Experimental Settings and Evaluation

We use the query-focused summarization task proposed in DUC[1] (2005-2007) to simulate our complex question answering experiments. We use the DUC-2006 data to train all the systems and then produce extract summaries for the DUC-2007 data. Supervised classifiers are typically trained on data pairs, defined by feature vectors and corresponding class labels. We use an automatic labeling approach to annotate the training data using ROUGE [1, 3, 9]. From each sentence of the training (and testing) data, we extract different query-related features and importance-oriented features such as: n-gram overlap, Longest Common Subsequence (LCS), Weighted LCS (WLCS), skip-bigram, exact word overlap, synonym overlap, hypernym/hyponym overlap, gloss overlap, Basic Element (BE) overlap, syntactic tree similarity measure, position of sentences, length of sentences, Named Entity (NE) match, cue word match and title match [1, 3, 5, 13].

For homogeneous ensemble, we divide the training data into 4 equal-sized fractions. Then, according to the CVC algorithm [2, 4, 11, 12], each time we leave separate 25% data out and use the rest 75% data for training. Thus, we generate 4 different SVM models. Next, we feed the test data to each of the generated SVM models which produces individual predictions (decision scores along with a label +1 or −1). The decision scores are the normalized distance from the separating hyperplane to each sample. To create the SVM ensemble, we combine the predictions by simple weighted averaging. We increment a particular classifier's decision value by 1 (giving more weight) if it predicts a sentence as positive and decrement by 1 (imposing penalty), if the case is opposite. The resulting prediction values are used later for ranking the sentences. During training steps, we use the third-order polynomial kernel for the SVM keeping the value of the trade-off parameter C as default. For our SVM experiments, we use the SVM^{light} package[2] [7]. The individual classifier settings for the heterogeneous ensemble formation are as follows. For SVM, we use the same setup as homogeneous ensemble. We implement the HMM model by Lin's HMM package[3]. We use the MALLET NLP toolkit [10] to implement the CRF. We modify its SimpleTagger class in order to include the provision for producing corresponding posterior probabilities of the predicted labels which were used later to rank the sentences. We build the MaxEnt system using Lin's MaxEnt package[4]. We combine the decision values of the four different classifiers by a weighted voting to build an ensemble. We impose a positive weight (ranging from 1 to 5 depending on the individual classifier's performance, more weight if it is declared positive by a better performer based on scores) to each positively classified sentence. We take no action for the negatively classified sentences so that they could fall back during ranking. The combined weighted votes of all the classifiers are used to rank the sentences to produce 250-word summaries [1].

---

[1] http://duc.nist.gov/

[2] http://svmlight.joachims.org/

[3] http://www.cs.ualberta.ca/~lindek/hmm.htm

[4] http://www.cs.ualberta.ca/~lindek/downloads.htm

We consider the multiple "reference summaries" of DUC-2007 to automatically evaluate our summaries using the ROUGE toolkit [9]. We compare the ensemble systems' performance with a baseline system. The baseline system's approach is to select the lead sentences (up to 250 words) from each topic's document set. In table 1, we present the ROUGE F-scores of different systems. We can see that the homogeneous ensemble improves the ROUGE-1, ROUGE-2 and ROUGE-SU scores over the baseline system by 16.2%, 26.6% and 30.3% respectively. The heterogeneous ensemble improves the ROUGE-1, ROUGE-2 and ROUGE-SU scores over the baseline system by 18.3%, 37.5% and 36.6% and over the homogeneous system by 1.80%, 8.64% and 4.79% respectively. Three native English speaking university graduate students judged[5] all the system generated summaries for readability (fluency) and overall responsiveness according to the TAC 2010 summary evaluation guidelines[6]. Table 2 presents the average readability and overall responsive scores of all the systems. The results again show that the ensemble systems perform better than the baseline system and heterogeneous ensemble performs the best in terms of overall responsiveness.

Table 1. ROUGE F-Scores for different systems

Systems	ROUGE-1	ROUGE-2	ROUGE-SU
Baseline	0.334	0.064	0.112
Homogeneous	0.388	0.081	0.146
Heterogeneous	0.395	0.088	0.153

Table 2. Readability and overall responsiveness scores for all systems

Systems	Readability	Overall Responsiveness
Baseline	4.24	1.80
Homogeneous	3.41	3.30
Heterogeneous	3.85	3.63

3   Conclusion and Future Work

In this paper, we presented the use of two ensemble methods: homogeneous and heterogeneous to perform the complex question answering task. Our experiments suggested the following: (a) ensemble methods outperform the conventional systems, and (b) heterogeneous ensemble performs the best for this problem. Future work is foreseen to use different learning algorithms for homogeneous ensemble and to improve the base classifiers' performance for both ensemble methods.

[5] The inter-annotator agreement of Fleiss' $\kappa = 0.63$ is computed for the three judges indicating a substantial degree of agreement [8].

[6] http://www.nist.gov/tac/2010/Summarization/
   Guided-Summ.2010.guidelines.html

Acknowledgments. The research reported in this paper was conducted at the University of Lethbridge and supported by the Natural Sciences and Engineering Research Council (NSERC) of Canada-discovery grant, and the University of Lethbridge.

References

1. Chali, Y., Hasan, S.A.: Query-focused Multi-document Summarization: Automatic Data Annotations and Supervised Learning Approaches. Journal of Natural Language Engineering 18(1), 109–145 (2012)
2. Chali, Y., Hasan, S.A., Joty, S.R.: A SVM-Based Ensemble Approach to Multi-Document Summarization. In: Gao, Y., Japkowicz, N. (eds.) Canadian AI 2009. LNCS (LNAI), vol. 5549, pp. 199–202. Springer, Heidelberg (2009)
3. Chali, Y., Hasan, S.A., Joty, S.R.: Do Automatic Annotation Techniques Have Any Impact on Supervised Complex Question Answering? In: Proceedings of the Joint conference of the 47th Annual Meeting of the Association for Computational Linguistics (ACL-IJCNLP 2009), Suntec, Singapore, pp. 329–332 (2009)
4. Dietterich, T.G.: Ensemble methods in machine learning. In: Kittler, J., Roli, F. (eds.) MCS 2000. LNCS, vol. 1857, pp. 1–15. Springer, Heidelberg (2000)
5. Edmundson, H.P.: New methods in automatic extracting. Journal of the ACM 16(2), 264–285 (1969)
6. Gashler, M., Giraud-Carrier, C.G., Martinez, T.R.: Decision Tree Ensemble: Small Heterogeneous Is Better Than Large Homogeneous. In: ICMLA, pp. 900–905 (2008)
7. Joachims, T.: Making large-Scale SVM Learning Practical. In: Advances in Kernel Methods - Support Vector Learning (1999)
8. Landis, J.R., Koch, G.G.: The Measurement of Observer Agreement for Categorical Data. Biometrics 33(1), 159–174 (1977)
9. Lin, C.Y.: ROUGE: A Package for Automatic Evaluation of Summaries. In: Proceedings of Workshop on Text Summarization Branches Out, Post-Conference Workshop of Association for Computational Linguistics, Barcelona, Spain, pp. 74–81 (2004)
10. McCallum, A.K.: MALLET: A Machine Learning for Language Toolkit (2002)
11. Parmanto, B., Munro, P.W., Doyle, H.R.: Improving committee diagnosis with resampling techniques. In: Advances in Neural Information Processing Systems, vol. 8, pp. 882–888 (1996)
12. Rooney, N., Patterson, D.W., Anand, S.S., Tsymbal, A.: Random subspacing for regression ensembles. In: FLAIRS Conference (2004)
13. Sekine, S., Nobata, C.A.: Sentence extraction with information extraction technique. In: Proceedings of the Document Understanding Conference (2001)
14. Silva, C., Ribeiro, B.: Rare class text categorization with SVM ensemble. Journal of Electrotechnical Review (Przeglad Elektrotechniczny) 1, 28–31 (2006)

Towards the Design of User Friendly Search Engines for Software Projects

Rafaila Grigoriou and Andreas L. Symeonidis

Electrical and Computer Engineering Dept., Aristotle University of Thessaloniki,
Thessaloniki, Greece
rafaila@ee.auth.gr, asymeon@eng.auth.gr

Abstract. Current work proposes a linguistic approach for supporting the identification of User requirements and Software Specifications. We introduce an NLP-based tool, PYTHIA, that serves as a search engine capable of handling software engineering terminology, aiming to close the loop between the end-user and the software developer. It is an ontology-based question answering system that employs semantic analysis as well as external (both generic use and domain-specific) dictionaries in order to handle term disambiguation, as posed in user defined queries.

1 Introduction

When developers get down to sketching and designing software, they are equipped with too few tools that could enable reuse of the optimal set of functional requirements and well engineered software modules satisfying these requirements. Were such tools available and this information properly stored, developers would be able to access other Software Engineers' solutions to similar projects and could reuse them as off-the-shelf components, or could adjust them to their own needs.

Taking this argument one step further, one could argue that such a "search-engine" for software projects should be interactive, allowing users to progressively identify the required software constructs, and adaptable, in order to increase its knowledge base. Question Answering (QA) systems could provide the means to realize such search engines, given that they illustrate these types of features.

Towards this direction we have designed and developed PYTHIA (**P**rogrammer's d**Y**namic **TH**ematic **I**nteractive **A**dvisor), a QA tool that can provide guidance through requirements elicitation and design specifications of a software project. Information related to already implemented software projects is stored in an especially-designed ontology scheme and offers engineers the ability to access previous design paradigms, reuse, or even evolve them.

2 Related Work

QA systems can be classified in two major categories, the open-domain and the restricted-domain systems [4]. The former may provide information on any

E. Métais, M. Roche, and M. Teisseire (Eds.): NLDB 2014, LNCS 8455, pp. 164–167, 2014.

topic/domain using any information they can access (on the web or knowledge bases), while the latter answer queries on specific topics/domains, using a focused set of information sources. The incorporation of semantics in QA systems can be performed in various stages of the query processing/query answering process, thus leading to four categories of systems: Semantics-based, Inference-based, Logic-based and Hybrid QA systems.

PYTHIA is an ontology-based, restricted-domain application and falls into the category of hybrid QA systems. Apart from Apple's Siri engine [5], one should also mention AQUA[3], which combines ontologies, logic and NLP technologies to exploit semantically annotated web pages and MOQA[1], which attempts to answer sequences of questions by using a fact repository comprising instances of ontological concepts. AQUAINT[6] employs a model-based approach, based on the idea that relations relevant to a question are best captured by an expressive model of events, while Unger and Cimiano [7] perform compositional meaning construction on the Semantic Web. Finally, Freya[2] combines syntactic parsing with a set of heuristic rules, as well as ontology-based annotations, in order to identify the answer type of input questions.

Although efficient, the above discussed systems cannot cope with the complexity and particularities of the software engineering process life-cycle. The use of a general-purpose lexicon was found inefficient, since most of engineering terms entail semantics and are not handled properly. PYTHIA users have the ability to perform multi-level queries, each one refining search, while they can also enrich the system knowledge base, in case the semantics of a concept are not properly handled. Section 3 discusses the main architectural components of PYTHIA.

3   System Architecture

3.1   PYTHIA Modules

PYTHIA[1] is a web-application that allows users to perform queries either in natural language, or by compiling advanced queries through the corresponding web pages. In both cases, the system is able to deal with term disambiguation with the help of external dictionaries. Information in PYTHIA is stored to and retrieved from a two-ontology scheme[2]: a) *RequirementsOnt*, where functional requirements are represented and stored in an actor-action-system-constraints form, and b) *UMLOnt*, which stores information on UML class diagrams and metadata defining the relation between entities. *RequirementsOnt* and *UMLOnt* are related through references on functional requirements.

PYTHIA comprises the following subsystems, which interact with each other during the flow of a query:

Spellchecking Subsystem - SPS: It provides similar functionality with Google's *did-you-mean* feature, checking whether the input text is orthographically correct. If not, the user is prompted to select from a number of suggested words,

---

[1] Available at: http://155.207.18.187:8080/pythia

[2] More information at: http://issel.ee.auth.gr/software-algorithms/

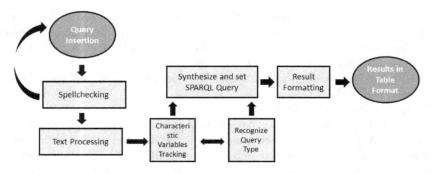

Fig. 1. System workflow diagram

ranked by similarity to the user input. SPS employs the Lucene Spellchecker API.

Text Processing Subsystem - TPS: TPS performs grammar and syntax analysis on the user input. It employs the Stanford Parser API in order to recognize 14 categories of dependencies, useful for the identification of the subject, the verb and the object or copula, as well as prepositional modifiers, noun compound modifiers and adjectival modifiers. Special cases such as passive voice, phrasal verbs and conjunctions are also well-handled.

Characteristic Variables Tracking Subsystem - CVTS: CVTS implements a Lexical Similarity Calculating Algorithm in order to identify the corresponding labels of the entities residing in the ontology, searching among class names and specific instances. CVTS analysis generates the subject, object, property and filter values that will be used for the creation of a SPARQL query to post to the semantic layer. Synonyms are also taken under consideration during the process, in order to increase the possibility of finding the correct relation, using WordNet.

Query Type Recognizing Subsystem - QTRS: QTRS interacts with CVTS, in order to identify the type of the query performed. Three types of queries can be performed: Simple, Dependency and Advanced Queries. Query categories and their properties are further discussed in next subsection. QTRS employs the Apache Jena API.

SPARQL Query Synthesis Subsystem - SQSS: SQSS composes the query based on the QTRS output and posts it to the ontology scheme.

Result Forming Subsystem - RFS: Finally, RFS is responsible for formatting results and generating the final user view. In all result representations, links to lower level information are provided.

3.2   Query Types Supported

PYTHIA supports 3 query categories: *Simple*, *Advanced* and *Dependency* queries. The user is confronted with a simple 2-tab interface, and is prompted to select between the *Simple* and *Advanced* query pane. In the former case, the user provides information in free text, while in the latter, the user can "guide" the query process by selecting the entities to be involved and probably specific query variables.

Simple queries are distinguished into the following subcategories: a) Simple queries in a Subject-Verb-Object (SVO) format, b) Subject-instance queries, c) Object-instance queries and, d) Dependency queries. Advanced queries can be of the following types: a) Advanced queries in SVO format, b) Object-instance advanced queries, c) Advanced queries with keywords instead of properties, iv) Dependency queries and, v) Step-by-step user created dependency queries.

If the subject and/or object entities are not identified as directly related to the ontology, queries are categorised as *Dependency* queries and multiple level SPARQL queries are built getting information from the user. Practically, when PYTHIA fails to retrieve any results related to a query, it provides the user with the capability to construct queries step-by-step, in a question answering manner. In all cases that the multiple level queries lead to results, the query is stored to an auxiliary database as a Dependency query. This way the system evolves automatically, enriching the variety of questions it is capable of answering.

4   Conclusions - Future Work

This paper describes the first approach towards the creation of a search engine for developers that want to retrieve information about past projects. PYTHIA is capable of replying to specific types of questions and learn from user queries. The presented methodology seems to be appropriate; however, specific actions could improve system's performance. User feedback could provide a useful means for quality improvement. Additionally, the generation of a software engineering lexicon and the replacement of WordNet, could also lead to better search results. Finally, keeping an archive of inserted questions and their results, as most search engines do nowadays, could help the system achieving quick and accurate responses to the most common queries.

References

1. Beale, S., Lavoie, B., McShane, M., Nirenburg, S., Korelsky, T.: Question answering using ontological semantics. In: Proceedings of the 2nd Workshop on Text Meaning and Interpretation, pp. 41–48 (2004)
2. Damljanovic, D., Agatonovic, M., Cunningham, H.: Identification of the question focus: Combining syntactic analysis and ontology-based lookup through the user interaction. In: LREC. Citeseer (2010)
3. Enrico, M.V.-V., Motta, E., Domingue, J.: Aqua: An ontology-driven question answering system, Stanford University. Citeseer (2003)
4. Merkel, A.: Using language models in question answering (2008)
5. Naone, E.: Tr10: Intelligent software assistant (March-April 2009)
6. Sinha, S., Narayanan, S.: Model-based answer selection. In: Proceedings of the AAAI Workshop on Inference for Textual Question Answering (2005)
7. Unger, C., Cimiano, P.: Pythia: Compositional meaning construction for ontology-based question answering on the semantic web. In: Muñoz, R., Montoyo, A., Métais, E. (eds.) NLDB 2011. LNCS, vol. 6716, pp. 153–160. Springer, Heidelberg (2011)

Towards a New Standard Arabic Test Collection for Mono- and Cross-Language Information Retrieval

Oussama Ben Khiroun[1], Raja Ayed[1], Bilel Elayeb[1,3], Ibrahim Bounhas[2], Narjès Bellamine Ben Saoud[1,4], and Fabrice Evrard[5]

[1] RIADI Research Laboratory, ENSI Manouba University 2010, Tunisia
{oussama.ben.khiroun,ayed.raja}@gmail.com,
Bilel.Elayeb@riadi.rnu.tn, Narjes.Bellamine@ensi.rnu.tn
[2] LISI Lab. of Computer Science for Industrial Systems, ISD Manouba University, 2010 Tunisia
Bounhas.Ibrahim@yahoo.fr
[3] Emirates College of Technology, P.O. Box: 41009, Abu Dhabi, United Arab Emirates
[4] Higher Institute of Informatics (ISI), Tunis El Manar University, 1002 Tunisia
[5] Informatics Research Institute of Toulouse (IRIT), 02 Rue Camichel, 31071 Toulouse, France
Fabrice.Evrard@enseeiht.fr

Abstract. We propose in this paper a new standard Arabic test collection for mono- and cross-language Information Retrieval (CLIR). To do this, we exploit the "Hadith" texts and we provide a portal for sampling and evaluation of Hadiths' results listed in both Arabic and English versions. The new called "Kunuz" standard Arabic test collection will promote and restart the development of Arabic mono retrieval and CLIR systems blocked since the earlier TREC-2001 and TREC-2002 editions.

Keywords: Mono- and Cross-Language Information Retrieval, Arabic Language, Standard Test Collection, Sampling Evaluation.

1 Introduction

The Arabic language contains 28 letters and uses right to left scripts for writing. It is considered to be morphologically rich because several words have the same orthographic form [1]. Arabic language presents many particularities since the lack of short vowels in many Arabic texts and the absence of capital letters which present many challenges in Information Retrieval (IR), Natural Language Processing [2], Word Sense Disambiguation and Named Entity Recognition [3] tasks.

The state-of-the art conferences which aim to build IR standard corpora like TREC[1], CLEF[2] and INEX[3] considered rarely Arabic language. Text REtrieval Conferences (TREC) held by NIST have greatly contributed to the construction of large and reliable test collections. TREC introduced a track for Arabic language

---

[1] http://trec.nist.gov/
[2] http://www.clef-initiative.eu/
[3] http://www.inex.otago.ac.nz/

E. Métais, M. Roche, and M. Teisseire (Eds.): NLDB 2014, LNCS 8455, pp. 168–171, 2014.

retrieval for the first time in 2001[4]. Indeed, the collection consists of a Cross-Language Information Retrieval (CLIR) track testing the use of English queries with regard to Arabic documents, as well as monolingual retrieval using Arabic queries. However, TREC-2001 and TREC-2002 Arabic texts did not include short vowels and focus on modern standard Arabic texts issued from newswire stories.

Thus, we feel the need of new vocalized corpora for traditional Arabic. The general methodology proposes to: (i) define a set of typical queries; (ii) experiment and combine as much as possible different analysis, indexing and matching models; (iii) aggregate the results of these models; and (iv) assess manually the results of these models.

This paper is organized as follows. First, we introduce the document collection and the test queries (topics) used to build the test collection respectively in Section 2 and Section 3. We present in Section 4 the relevance judgment process. Finally, we conclude this paper in Section 5.

2   Documents Collection

Hadith related literature exists in several formats and types of documents/databases. Different versions having different sources and content are available. Thus, the input of our system should be studied and verified carefully. Besides, any of these versions has its limits and advantages and it is hard to choose the best one, especially that none of the existent versions is multilingual nor the hadiths are well aligned. We exploit the Hadith corpus because it is big, vocalized and contains diversified knowledge to develop approaches and models for knowledge extraction and information retrieval [5].

```
<DOC>
  <DOCNO>B-001-001-001</DOCNO>
  <DOCID>B-001-001-001</DOCID>
  <GENRE>بَدْءُ الْوَحْيِ</GENRE>
  <SUBJECTS>بَدْءُ الْوَحْيِ</SUBJECTS>
  <NAMES>قَالَ عُمَرَ بْنَ الْخَطَّابِ</NAMES>
  <TEXT>
سَمِعْتُ رَسُولَ اللَّهِ صَلَّى اللَّهُ عَلَيْهِ وَسَلَّمَ ، يَقُولُ : " إِنَّمَا الْأَعْمَالُ بِالنِّيَّاتِ ، وَإِنَّمَا لِكُلِّ امْرِئٍ مَا نَوَى ، فَمَنْ كَانَتْ هِجْرَتُهُ إِلَى دُنْيَا يُصِيبُهَا
أَوْ إِلَى امْرَأَةٍ يَنْكِحُهَا ، فَهِجْرَتُهُ إِلَى مَا هَاجَرَ إِلَيْهِ" .
  </TEXT>
</DOC>
```

Fig. 1. A sample document of "Kunuz" collection (Arabic version)

Figure 1 presents an Arabic sample of the Hadiths' document collection structured in the XML TREC format (an English version of each hadith exists also in the collection respecting the same XML structure).The Hadith collection contains more than 6.700 texts having an average of 70 words per Hadith.

The major number of Hadiths' texts counts less than 1000 words and about 80% of Hadiths have less than 100 words. Thus, we can consider that Hadiths' texts in "Kunuz" collection could be considered as short documents.

3   Topics Selection

The set of 58 standard test queries (topics) in "Kunuz" collection have been selected from suggested "fatwas" (فتوى) by users on *Islamweb* website[4]. Indeed, *Islamweb* is a site of Islamic preaching which is run by an elite group of graduates in theology, literature and various technological fields. Both question and answer parts of fatwas was used in title and description tags respecting the XML TREC format. An overview of "Kunuz" topics is listed in Table 1.

Table 1. Overview of topics (test queries) in "Kunuz"collection

Property	Value
Number of topics	58
Total number of words in title section	306 words
Average length of title section	~ 5 words per title section
Total number of words in description section	1069 words
Average length of description section	~ 18 words per description section

4   Relevance Judgment

According to the state of the art campaigns for building reference test collections, the relevance judgment task is considered critical. Actually, it is feasible to carry out complete relevance judgment on small test collections for all document-query pairs. However, in large test collections such the Hadith, this approach became difficult to accomplish. Consequently, we based the judgment process on the recommended TREC pooling system methodology [6]. For this task, we varied six Arabic stemming tools: N-gram, Khoja, AraMorph, Al-Stem-Darwish, Al-Stem-Alex [1] and Possibilistic morphological analyzer [7] and three matching models (BM25, PL2 and DFRee) specified in Terrier[5] IR framework [8].

Finally, the created pools are manually examined by human users to create the list of correct relevant documents. So, we developed a Web portal named "Kunuz Al Mustafa" (كنوزالمصطفى) to facilitate the sampling and the evaluation tasks of the Hadith standard test collection.

5   Conclusion and Future Works

In this paper, we introduced "Kunuz", the new test collection for Arabic mono- and cross-language information retrieval by detailing the construction process. The "Kunuz" project still needs collaborations, in both human resources and hosting funding help, to accomplish the relevance judgment task through the "Kunuz Al Mustafa" Web portal.

---

[4] http://fatwa.islamweb.net/fatwa/index.php
[5] http://terrier.org/

Transforming the "Hadith" corpus into a new standard test for Arabic IR and CLIR is a promising project and the construction approach is not restricted only to religious texts. The goal is to boost research works on especially Arabic CLIR blocked since TREC-2001 & TREC-2002, and consequently provide an opportunity for new Arabic CLIR competitions via call for workshops, conferences or journals papers using this standard. However, there are some challenges to deal with such that (i) The size of the hadith corpus and the manual assessment of results; and (ii) The missing data and translations. On the one hand, the hadith corpus is huge counting thousands of hadiths distributed in heterogeneous versions in several languages. All this data should be aligned and rechecked manually. On the other hand, while aligning the first book, we remarked that some hadiths are not translated. In other cases, a hadith has an English version, but some of its words are not well translated; many of them are just transliterated, which may challenge IR systems. According to these difficulties and given the great efforts to be performed to convince the IR community about the utility and the quality of this resource, we estimate that the hadith corpus needs a great effort and a hard work to be standardized in the next few months.

References

1. Abu El-Khair, I.: Arabic information retrieval. Annu. Rev. Inf. Sci. Technol. 41, 505–533 (2007)
2. Beseiso, M., Ahmad, A.R., Ismail, R.: A Survey of Arabic language Support in Semantic web. Int. J. Comput. Appl. 9, 35–40 (2010)
3. Zayed, O., El-Beltagy, S., Haggag, O.: An Approach for Extracting and Disambiguating Arabic Persons' Names Using Clustered Dictionaries and Scored Patterns. In: Métais, E., Meziane, F., Saraee, M., Sugumaran, V., Vadera, S. (eds.) NLDB 2013. LNCS, vol. 7934, pp. 201–212. Springer, Heidelberg (2013)
4. Gey, F.C., Oard, D.W.: The TREC-2001 Cross-Language Information Retrieval Track: Searching Arabic Using English, French or Arabic Queries. In: The Tenth Text REtrieval Conference (TREC), pp. 16–25 (2002)
5. Bounhas, I., Elayeb, B., Evrard, F., Slimani, Y.: Toward a Computer Study of the Reliability of Arabic Stories. J. Am. Soc. Inf. Sci. Technol. 61, 1686–1705 (2010)
6. Clarke, C.L.A., Craswell, N., Soboroff, I., Cormack, G.V.: Overview of the TREC 2010 Web Track. In: The 19th Text REtrieval Conference (TREC) (2011)
7. Ayed, R., Bounhas, I., Elayeb, B., Evrard, F., Bellamine Ben Saoud, N.: Arabic Morphological Analysis and Disambiguation Using a Possibilistic Classifier. In: Huang, D.-S., Ma, J., Jo, K.-H., Gromiha, M.M. (eds.) ICIC 2012. LNCS, vol. 7390, pp. 274–279. Springer, Heidelberg (2012)
8. Ounis, I., Amati, G., Plachouras, V., He, B., Macdonald, C., Lioma, C.: Terrier: A High Performance and Scalable Information Retrieval Platform. In: Proceedings of ACM SIGIR 2006 Workshop on Open Source Information Retrieval (OSIR), pp. 18–25 (2006)

Sentiment Analysis Techniques for Positive Language Development

Izaskun Fernandez[1], Yolanda Lekuona[2], Ruben Ferreira[1], Santiago Fernández[1], and Aitor Arnaiz[1]

[1] Ik4-Tekniker
Polo Ténologico de Eibar, C/ Iñaki Goenaga, 5 - 20600 Eibar-Gipuzkoa, Spain
{izaskun.fernandez,ruben.ferreira,
santiago.fernandez,aitor.arnaiz}@tekniker.es
[2] Mondragon Corporación - Otalora
B^o Aozaraza n^o2. 20550-Aretxabaleta-Gipuzkoa, Spain
ylekuona@mondragoncorporation.com

Abstract. With the growing availability and popularity of opinion-rich resources such as on-line review sites and personal blogs, the use of information technologies to seek out and understand the opinions of others has increased significantly. This paper presents Posimed, a sentiment assessment approach that focuses on verbal language using information technologies for Spanish. We describe how Posimed combines natural language technologies for Spanish and expert domain knowledge to extract relevant sentiment and attitude information units from conversations between people (from interviews, coaching sessions, etc.) and supports the programmes that positivity training experts provide in order to develop the *Positivity competence*. We have evaluated Posimed both in a quantitative and a qualitative way and these evaluations show that Posimed provides an accurate analysis (73%) and reduces significantly (80% reduction) the time for the same job when it is performed manually by the domain expert.

Keywords: Natural Language Processing, Rule-based, Positive Communication, Sentiment Analysis and Assessment, Mindset.

1 Introduction

Positivity can be defined as the positive attitude that can be observed through a person's behaviour and language, both verbal and non-verbal. According to current trends in research and the application of positive psychology [1], positive thinking and positive language can be developed in order to improve the positive attitude. For that purpose, a comprehensive (thinking, emotion, and body language) training and coaching methodology is necessary.

The evaluation of the results of this training is directly related to the individual's physical and verbal expression. Considering sentiment analysis, the field that determines the attitude of a speaker or a writer through natural

E. Métais, M. Roche, and M. Teisseire (Eds.): NLDB 2014, LNCS 8455, pp. 172–183, 2014.

language processing with respect to some topic or polarity of a document, it seems clear that at least the verbal analysis is supported, which comprises the most significant part of the evaluation.

The rise of social media such as blogs and social networks considered as opinion-rich resources has considerably increased the interest in sentiment analysis. However, few works on information technologies studying the implicit opinion and sentiment in communication between people can be found.

In this paper we describe our approach on developing Posimed, a Spanish support tool for positive communication evaluation combining natural language processing techniques and rules (representing domain expert knowledge). Posimed aims to automate the process of analysing Spanish expressions that denote positivity/negativity as much as possible, and it gives support to reporting of the communication tendency, attitude and mindset of the person in terms of positivity.

The evaluation report will be the starting point for applying the training and coaching methodology mentioned before to an individual, and also the key for an efficient and cost-effective evaluation of the individual's evolution, which is the main current issue.

The positivity evaluation report is a key tool for developing the positivity competence and it works together with the appropriate training methodology. It is used as the starting point and also to measure the person's improvement during and after the process. Posimed allows the report to be created in a very time and cost-effective way thanks to its automatic mode, which is the main benefit for domain experts.

2 Background

According to Damasio [2], the positive language activates various psychological mechanisms of emotional regulation such as re-evaluation, re-formulation or re-interpretation. These mechanisms seem to give the chance to redirect the automatic emotional reaction to a more positive or tolerable one.

Positive language ability to regulate emotions depends on the affective style of each person [3], and we can change the way we feel and change our thoughts by changing internal language [4]. The most common way is to use positive language (words or phrases) that induce pleasant feelings and thoughts in order to re-evaluate or re-interpret the negative reactions. From a methodological and theoretical point of view, the main motivation of our project has been to provide a comprehensive methodology to develop this positive language, with the Posimed tool, the technical approach presented in this paper, as a part of it. Specifically, in this paper we focus on showing how a sentiment analysis approach, based on natural language processing techniques combined with domain expert knowledge through rules, can help in such a positive language development evaluation.

Research on sentiment analysis shows different approaches to automatically predict the sentiments of words, expressions and/or documents. These approaches are Natural Language Processing (NLP) and pattern-based techniques, supervised and unsupervised machine learning approaches, and finally hybrid approaches.

There are many publications in this area that apply these different approaches. For example, if we focus on the SemEval-2013 Task 2 (Sentiment Analysis in Twitter) [5], we can observe that most of the systems are supervised machine learning approaches (8 system use SVM, 7 Naive Bayes and 3 Maximum Entropy), including the winning one [6] in most of the evaluated subtasks. But there is also a pattern-based system [7], which despite not winning, is very close to the best system performance that relies on an existing Domain Independent sentiment taxonomy for English and hand-written rules for sentiment phrases detection. So a rule-based approach can be a good approximation and its benefit is that no training data is needed for its construction. And besides, it could be an interesting approach for annotated corpora generation.

As mentioned before, there are also approaches that combine both kinds of systems as described in [8]. More specifically in this work, the authors describe how the SVM trained model outperforms the rule-based model, and how they are complementary. Used together the results can be improved.

For Posimed, we have selected the rule-based approach for the following reasons: an annotated corpus is not necessary; it is possible to explicitly exploit the knowledge of domain experts; it provides a strategy to semi-automatically generate an annotated corpus; it is domain/context independent just like the rules are; and finally, it fulfils the domain experts requirement of the capability to adapt/model existing/new rules for new scenarios without any resource dependency, apart from their knowledge.

3   Solution

Posimed is a web-based tool which automatically analyses the positivity/negativity of a text and also provides a manual mode that allows domain experts to make their own corrections and annotations. Given a plain text in Spanish (from an interview, coaching session, article, speech, etc.) Posimed automatically identifies and classifies expressions denoting sentiment, attitude and mindset (in terms of positivity/negativity) in order to generate a positivity evaluation report.

Manual mode lets the domain experts modify the Posimed annotation results by adding, deleting and/or changing using the Posimed GUI. After optional manual edition, they can ask for the statistical positivity report to be automatically generated (see Fig. 1).

So Posimed is not only a tool for automatically annotating sentiment and attitude expressions but also it can be considered an environment for improving the automatic results and for creating experts' supervised positivity evaluation reports. The following subsections give a detailed description of automatic sentiment and attitude analysis annotation and the GUI for final user interactions.

Fig. 1. Example report section

3.1 Architecture

We propose a service-oriented architecture for the Posimed web tool. As it is shown in Fig. 2, there are two main services for extracting the relevant information related to positivity from a plain text: the annotation service and the sentiment analysis service.

Annotation Service. The annotation service provides a specialized terminology-based morphosyntactic annotation. Given a plain text, the annotation service generates an output where each (multi)word is described by its morphosyntactic information and also by a special class, the class of the specialized terminology it belongs to, if applicable.

The service is based on two main resources: Freeling[1] for morphosyntactic annotation purposes and a specialized terminology resource for providing sentimental information at multi(word) level. Combining both resources as described below, the service gets the necessary information for the next step: the extraction of sentiment expressions.

Freeling[9] is an open-source multilingual language processing library providing a wide range of analysis functionalities for several languages. In the Posimed annotation service we use the Spanish lemmatizer, the part of speech tagger, the

---

[1] http://nlp.lsi.upc.edu/freeling/

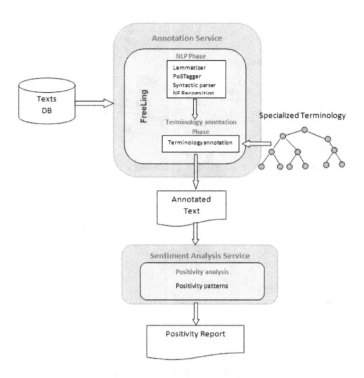

Fig. 2. Posimed Architecture: Main Services

syntactic analyser and the NE Recognition functionalities for basic morphosyntactic annotation and the multiword annotation functionality for integrating the
specialized terminology in Freeling. This last functionality takes into account
the (multi)words and their corresponding class defined at the specialized terminology resource for generating the final morphosyntactic output: an output that
contains a morphosyntactic tag (PoS Tag) and an optional sentimental class for
each (multi)word defined at the specialized terminology, if applicable as shown
in Fig. 3.

For the specialized terminology definition, a group of domain experts have
used their knowledge of the Spanish language together with data coming from
a collection of 37 interviews, 12 coaching sessions and 30 answers to a certain
query. They have reviewed the text collection identifying (multi)words with sentiment connotation and dividing them into in different classes. These classes
are sentiment-related classes as shown in Fig. 4 with a small set of classes and
examples.

Some classes are generics such as positive or negative adverbs, verbs or nouns.
But there are also some more detailed ones such as verbs denoting actions, opinions or agreement, denoting positive/negative actions, positive/negative judge-
ments, trust, engagement, gratitude, opportunity, and so on that are considered

```
<?xml version='1.0' encoding='utf-8'?>
<analysis>
  <sentence>
    <token id='1' literal='Esto' word='esto' lema='este' post='PD0NS000'>Esto</token>
    <token id='2' literal='se' word='se' lema='se' post='P00CN000'>se</token>
    <token id='3' literal='me' word='me' lema='me' post='PP1CS000'>me</token>
    <token id='4' literal='da' word='da' lema='dar' post='VMIP3S0#VER_APO
    #VER_JUI#VER_ACC_POS#VER_DSR#VER_CON'>da</token>
    <token id='5' literal='peor' word='peor' lema='peor' post='AQ0CS0#ADV_EST_NEG
    #ADJ_CAN_INT_RED#ADJ_NEG#ADJ_NEG_NEG#SUS_NEG#SUB_NEG#ADV_MOD_NEG#ADV_MOD_NEG_NEG'>peor</token>
    <token id='6' literal='en' word='en' lema='en' post='SPS00'>en</token>
    <token id='7' literal='este' word='este' lema='este' post='DD0MS0'>este</token>
    <token id='8' literal='idioma' word='idioma' lema='idioma' post='NCMS000'>idioma</token>
  </sentence>
</analysis>
```

Fig. 3. Example of the annotation service output

relevant for identifying expressions concerning to positivity competence. In any case, they are classes that aim to provide additional and relevant information for sentiment and attitude expression recognition that morphosyntactic tags do not provide.

VERBO JUICIOS	ADJETIVO POSITIVO	ADJETIVO NEGATIVO	VERBO DE ACCION POSITIVA	VERBO DE ACCION NEGATIVA	ADJETIVO DE EMOCIÓN agradable	ADJETIVO DE EMOCIÓN desagradable	ADJETIVO de EMOCION PODER
VER_JUI	ADJ_POS_POS	ADJ_NEG_NEG	VER_ACC_POS_POS	VER_ACC_NEG_NEG	ADJ_EMO_AGR	ADJ_EMO_DES	ADJ_POD
andar	a gusto	abatido	abrir camino	Abandonar	a gusto	apático	a la altura
causar	accesible	Aborrecible	acoger	abortar	acogedor	a medio morir saltando	accesible
cesar	acertado	aborrrecido	Aconsejar	Aburrir	activado	abandonado	Activo
cobrar	acogedor	abucheado	Adaptar	abusar	activo	abatido	asequible
coger	adecuado	aburrido	afrontar	Acobardar	admirable	aborrecible	capacitado
conservar	Admirable	abusador	agradecer	acosar	admirado	abúlico	Capaz
considerar	admirado	achuchado	alegrar	agredir	adorable	aburrido	cómodo
creer	admisible	acobardado	alentar	ahorcar	adorado	acongojado	competente
dar	adorable	acongojado	amar	Aislar	afanoso	afanado	confiado
embargar	adorado	Acorralado	animar	amenazar	afectado	afectado	consciente
encontrar	afable	Acosado	apagar fuegos	angustiar	afectuoso	afligido	Convencido
estar	afectuoso	acosador	aportar	arruinar	afortunado	agitado	Creador
estar siendo	afín	Acusado	apoyar	Asesinar	agradable	agobiado	creativo
experimentar	afinado	afligido	aprender	atacar	agradecido	agónico	directo
fomentar	afortunado	agobiado	asesorar	atascar	alegre	agotado	Eficaz
ganar	agradable	agónico	asumir	Bloquear	aliviado	agresivo	Emprendedor

Fig. 4. Examples of some specialized terminology classes

There are a total of 325 classes with their corresponding (multi)word list that comprises all the specialized terminology. Terminology which, as we have mentioned before, we have used for adapting Freeling output for our needs (mainly through the multiword identification functionality).

Sentiment Analysis Service. Given a specialized terminology-based morphosyntactically tagged text, the sentiment analysis service extracts the sentiment and attitude expressions in it. The key resource for that is a rule-set repository where the domain experts knowledge concerning to relevant expressions denoting communication profiles is represented. This rule-set repository consists of patterns describing potential expressions using the specialized terminology-based morphosyntactic tags.

For the rule-set repository construction, we have applied an iterative process in collaboration with domain experts. As a first step, experts have identified which communication profile types (families) are relevant when creating the positivity report manually. Next, for each profile type they have determined which are the kind (category) of patterns that describes it. And as a final step of the first iteration, they have defined the patterns for each category. In subsequent iterations, and based on the validation outputs of the patterns defined at previous iterations, experts have been working on including patterns and creating new categories and even profiles.

The patterns are defined in terms of (multi)word sequences, detailing the following information for each (multi)word:

$$"form" (lemma)[PoS\_Tag] < specialized\_terminology's\_category >$$

Except for the PoS tag element, it is possible to add a PoS tag specification using '/' in the description of any element.

It is not necessary to describe all the elements for each word, only the necessary or relevant ones (usually one with an optional PoS tag specification). For instance, in the following example we show a pattern that matches expressions like *le parece estupendo*[2], where the first element is described by the personal pronoun PoS Tag (*PRO_PER*), and the second and the third ones by their corresponding specialized terminology category (*VER_JUI, judgement verb* and *ADJ_POS, positive adjective*), but in the latter with an added specification that restricts the element to belong also to *ADJ(adjective)* PoS tag.

$$[PRO\_PER] < VER\_JUI >< ADJ\_POS/ADJ >$$

As we have previously mentioned, each pattern belongs to a category, so that when a sequence of (multi)words fits a pattern, the whole sequence is annotated as an expression and is classified in that category. When a (multi)word sequence satisfies more than one pattern in the repository, the Sentiment Analysis Service selects the longer one, avoiding nested annotations. At Fig. 5, you can see an example where a *negative judgement* is annotated between $< MW >$ tags.

```xml
<?xml version='1.0' encoding='utf-8'?>
<analysis>
  <sentence>
    <token id='1' literal='Esto' word='esto' lema='este' post='PD0NS000'>Esto</token>
    <token id='2' literal='se' word='se' lema='se' post='P00CN000'>se</token>
    <MW type='Juicio negativo'>
      <token id='3' literal='me' word='me' lema='me' post='PP1CS000'>me</token>
      <token id='4' literal='da' word='da' lema='dar' post='VMIP3S0#VER_APO
#VER_JUI#VER_ACC_POS#VER_DSR#VER_CON'>da</token>
      <token id='5' literal='peor' word='peor' lema='peor' post='AQ0CS0#ADV_EST_NEG
#ADJ_CAN_INT_RED#ADJ_NEG#ADJ_NEG_NEG#SUS_NEG#SUS_NEG_NEG#ADV_MOD_NEG#ADV_MOD_NEG_NEG'>peor</token>
      <token id='6' literal='en' word='en' lema='en' post='SPS00'>en</token>
      <token id='7' literal='este' word='este' lema='este' post='DD0MS0'>este</token>
      <token id='8' literal='idioma' word='idioma' lema='idioma' post='NCMS000'>idioma</token>
    </MW>
  </sentence>
</analysis>
```

Fig. 5. Example of the sentiment analysis service output

---

[2] It looks great.

Currently, the rule-set repository consists of 29.237 patterns[3] which are divided into 108 categories such as *positive vs. negative words, possibility judgement vs. impossibility judgement, responsible vs. victim attitude, pleasant vs. unpleasant emotions, empowered vs. weak inner dialogue,* and so on. Since these categories are related to communication profiles (actually 43 families) and Posimed automatically identifies and classifies expressions according to these patterns, we can say that Posimed automatically provides the necessary information for generating a communication profile report.

Report Generation. The last but not least important service is the report generation, which using the previous functionality outputs creates a statistical report summing up the tagged information and also including all the tagged expressions and manual changes performed by the domain expert. The report sums up the log of manual editing in an excel document, and for each category: the number of expressions tagged automatically; the number of manually edited expressions, detailing adds and deletes; and the expressions identified on the text.

3.2   GUI

As with every automatic process, the automatic annotation process is not 100% accurate and generates annotation errors. Lots of them are errors inherited from the Freeling process, where the output tags are not the expected ones (often due to structurally incorrect/incomplete sentences that we usually use in oral communication) and so they do not match the patterns, missing potentially relevant expressions. Sometimes it is a circumstantial error but not always. In the latter case, the patterns need to be revised using similar examples to reach a consensus between Freeling usual output and the patterns, and use it for some redefining of patterns. But when it is a circumstantial error, it is possible to solve it easily by manually modifying that particular Posimed automatic annotation.

To solve the last situation in an easy and interactive way, we have designed and developed a GUI[4] that gives the chance to experts to modify the automatic output: deleting, adding or modifying annotations just with mouse selections and right clicks. GUI has some more functionalities, all of them to facilitate the experts' interaction with the described services and configuration of text annotation. We can distinguish two main types of functionalities: functionalities regarding a text annotation process; and those more oriented to the system configuration.

The functionalities of a text annotation process include both, the settings definition for a particular text annotation and the post-edition.

– Selection of the categories to consider during the annotation process

---

[3] About half of the patterns are automatically generated based on the other half of manually defined patterns, in order to resolve mainly pronouns syntactic behaviours.

[4] http://posimed.tekniker.es/

Fig. 6. Posimed web tool: Interactive annotation

- Different visualizations for the automatic annotation output
- Output's visual edition for error correction

The configuration functionalities allow the tool administrator to load a new version of specialized terminology and rule-set. This way, the tool offers the required flexibility to adapt the tool to new scenarios or needs. It also includes a pattern tester for validation purposes if the new patterns are well-defined and are coherent with Freeling output.

The modular design of the tool and the functionality that keeps the key resources independent from the tool makes it a very flexible and configurable tool that can evolve in the future without any expert software development assistance. This is a very basic feature, since the tool is thought to be used in different contexts where the patterns can change and so the key resources are highly likely to change.

4   Evaluation

For the Posimed web tool evaluation we have considered two aspects: the tool accuracy (henceforth, quantitative evaluation) and the impact of the tool for time/cost/effort/manual work reduction (hereafter, qualitative evaluation). While for the latter evaluation is based partially on the experts' subjective impression, the former has been performed using the basic measures for information extraction and retrieval systems: precision and recall.

4.1   Quantitative Evaluation

In order to estimate the tool accuracy, we have used 10 real interviews of about 45 minutes each comprising a corpus of about 33,000 words. All the interviewed people were business managers that were involved in a positivity competence development process. After the interview, they receive their profile report. Posimed is used for the purpose of creating these reports, providing a good evaluation of the positivity competence profile of the person.

We have automatically processed those 10 interviews identifying and classifying the longest expressions that match any of the patterns defined. After the automatic tagging process, an expert has reviewed each interview deleting the incorrect assignments, adding new ones and also modifying the categories of any expressions that s/he has considered appropriate.

After this manual review we get an evaluation corpus with 12,172 relevant expressions, constructed in a semi-automatic way. But we have not only used the manual review for corpus and evaluation purposes. Experts have also used the identified errors for improving the patterns, and therefore the tool accuracy.

In Table 1, we show the results of the manually review work in terms of the number of expressions extracted by Posimed versus the final correct expression set, detailing the amount of deleted and added expressions.

Table 1. Posimed vs. Correct expressions

	Posimed	Manually added	Manually removed	Correct-set
Interview1	1,234	90	540	784
Interview2	2,042	155	769	1,428
Interview3	1,456	112	508	1,060
Interview4	1,396	174	482	1,088
Interview5	1,679	236	646	1,269
Interview6	1,164	84	401	847
Interview7	2,055	341	715	1,681
Interview8	1,517	146	543	1,120
Interview9	1,310	89	562	837
Interview10	2,408	318	668	2,058
Total	**16,261**	**1,745**	**5,834**	**12,172**

Analysing the results, we can observe that only 23% of the total corrections represent missing (added) annotations while 77% of the corrections are deletes. The deletes are mostly implemented as a result of expected 'false positives'. As the analysts prefer Posimed to generate 'false positives' rather than assume the risk of missing potential expressions, they have defined patterns knowing that these patterns will annotate expressions that don't correspond and they should be deleted on the automatic output manual review.

This decision does not benefit the Posimed tool at least in terms of precision as shown in Table 2.

Table 2. Posimed tool quantitative evaluation

Precision	Recall	F₁
64.12%	85.66%	73.34%

However, and possibly due to general behaviour patterns, Posimed recall is quite high and it balanced the overall scope of the tool obtaining quite a good performance (73.34%).

4.2 Qualitative Evaluation

For qualitative evaluation, we have used the experience of the domain experts annotating and generating positivity evaluation reports and we have compared it with the same process supported on Posimed, mainly in terms of time.

For an expert analyst, analysing a 45 minutes interview and generating the communication profile report without any automatic support takes approximately 20 hours. When using Posimed for that process (we have tested with the 10 interviews recently mentioned), the time is reduced to 4 hours, which means a reduction of 80% of the time required. This time reduction makes the process cheaper and more accessible not only for managers but also for more employees/people.

As we have mentioned before, we have defined the rule-set repository iteratively, validating the output of each iteration on a set of texts (about 37 interviews) belonging to different people and 12 coaching sessions. For this validation, at the beginning of the project an analyst analysed this text-set manually and after that, at each rule-set definition iteration, we compared both manual and automatic outputs in order to identify errors and improve the rule-set. In those comparisons we detected not only errors that help us to improve the rule-set and therefore Posimed performance, but also we showed that sometimes Posimed identified relevant expressions that the expert had ignored accidentally during manual tagging.

So the time and therefore the cost reduction are not the only benefit. According to analysts' experience and feedback, Posimed makes it easy to analyse texts interactively and the assurance of tagging expressions that the expert can accidentally overlook. Interestingly enough, Posimed increases human expert precision.

In general, we can say that the results show that Posimed is a useful resource to use for communication positivity evaluation.

5 Conclusions Further Work

We have presented the work that we have carried out in the field of sentiment analysis. More specifically, we have described the support tool that we have created to assist human analyst experts in the analysis of texts. The main goal of our tool is to extract positivity competence characteristics and report them through a profile definition.

We have shown that with an acceptable accuracy (73.34% of F_1), Posimed considerably improves manual process, increasing the human accuracy and reducing the time and cost of the process. As a result, positivity communication development can become an accessible methodology for more people.

In order to utilise all the knowledge generated in this work as much as possible, we plan to enlarge the manually reviewed corpus using the semi-automatic annotation approximation as we have done for evaluation purposes, and use it as a source for a machine-learning based system development. We are quite sure that the results will be improved by combining both strategies, reducing the manual work and therefore the process time and cost.

References

1. Seligman, M.E.P.: La auténtica felicidad. EDICIONES B, S.A (2003) ISBN 9788466611480
2. Damasio, A.: The feeling of what happens. Harcourt Brace, New York (1999)
3. Davidson, R.J.: The neuroscience of affective style. The New Cognitive Neurosciences 2, 1149–1159 (2000)
4. Ochsner, K.N., Gross, J.J.: Thinking makes it so: a social cognitive neuroscience approach to emotion regulation. In: Handbook of Self-Regulation: Research, Theory, and Applications, pp. 229–255 (2004)
5. Nakov, P., Kozareva, Z., Ritter, A., Rosenthal, S., Stoyanov, V., Wilson, T.: Semeval-2013 Task 2: Sentiment analysis in twitter. In: SemEval 2013, Atlanta, Georgia (2013)
6. Mohammad, S.M., Kiritchenko, S., Zhu, X.: NRC-Canada: Building the State-of-the-Art in Sentiment Analysis of Tweets. In: SemEval 2013, Atlanta, Georgia (2013)
7. Reckman, H., Baird, C., Crawford, J., Crowell, R., Micciulla, L., Sethi, S., Veress, F.: Teragram: Rule-based detection of sentiment phrases using SAS Sentiment Analysis. In: SemEval 2013, Atlanta, Georgia (2013)
8. Kawathekar, S.A., Kshirsagar, M.M.: Sentiments analysis using Hybrid Approach involving Rule-Based & Support Vector Machines methods. IOSRJEN 2(1), 55–58 (2012)
9. Padró, L., Stanilovsky, E.: FreeLing 3.0: Towards Wider Multilinguality. In: LREC 2012. ELRA, Istanbul (2012)

Exploiting Wikipedia for Entity Name Disambiguation in Tweets

Muhammad Atif Qureshi[1,2], Colm O'Riordan[1], and Gabriella Pasi[2]

[1] Computational Intelligence Research Group, Information Technology,
National University of Ireland, Galway, Ireland
[2] Information Retrieval Lab, Informatics, Systems and Communication,
University of Milan Bicocca, Milan, Italy
{muhammad.qureshi,colm.oriordan}@nuigalway.ie, pasi@disco.unimib.it

Abstract. Social media repositories serve as a significant source of evidence when extracting information related to the reputation of a particular entity (e.g., a particular politician, singer or company). Reputation management experts are in need of automated methods for mining the social media repositories (in particular Twitter) to monitor the reputation of a particular entity. A quite significant research challenge related to the above issue is to disambiguate tweets with respect to entity names. To address this issue in this paper we use "context phrases" in a tweet and Wikipedia disambiguated articles for a particular entity in a random forest classifier. Furthermore, we also utilize the concept of "relatedness" between tweet and entity using the Wikipedia category-article structure that captures the amount of discussion present inside a tweet related to an entity. The experimental evaluations show a significant improvement over the baseline and comparable performance with other systems representing strong performance given that we restrict ourselves to features extracted from Wikipedia.

1 Introduction

Companies are increasingly making use of social media for their promotion and marketing. At the same time social media users voice their opinions about various entities/brands (e.g., musicians, movies, companies) [5]. This has recently given birth to a new area within the marketing domain known as "online reputation management" where automated methods facilitate monitoring reputation of entities instead of relying completely on the manual reputation management by an expert as was done traditionally [1]. One significant subtask that serves as the building block of efforts within "online reputation management" is the task of "entity name disambiguation" which resolves any ambiguous entity into a particular entity of interest (e.g., distinguishing between the fruit "Apple" and the company "Apple"). Entity name disambiguation in tweets is a significant research challenge due to both the short text in tweets and the considerable amount of noise in them, and this paper is an attempt to solve this challenging subtask of online reputation management. We describe our experience in

E. Métais, M. Roche, and M. Teisseire (Eds.): NLDB 2014, LNCS 8455, pp. 184–195, 2014.

devising an automated algorithm for dealing with the "entity name disambiguation" challenge in tweets. Starting from promising results we obtained within the CLEF 2013 RepLab filtering task[1], we present an extension of our previously proposed technique [11] by taking advantage of the Wikipedia category-article structure in addition to the previously proposed Wikipedia articles' hyperlink structure[2].

The strength of our approach comes from the exploitation of the encyclopaedic knowledge in Wikipedia; all participants of the RepLab filtering task that relied mainly on Wikipedia for the CLEF RepLab filtering subtask have failed to show competitive results, e.g. [10] which used an established system defined by Meij et al. [8]. On the other hand, our technique is the only technique which is solely based on Wikipedia that shows a significant performance over the baseline while at the same time showing comparable performance over other methods proposed by all other systems that participated in the CLEF 2013 RepLab filtering subtask. The novelty of our technique lies in utilizing just the content available in a tweet and Wikipedia while the other teams made use of several rich features such as url mentions inside a tweet, usernames inside a tweet, official websites of entities, linguistic processing rules etc. Our technique makes use of the knowledge encoded in the Wikipedia graph structure for entity name disambiguation in tweets. We utilize the Wikipedia disambiguation pages for an entity to determine the amount of disambiguation within a particular tweet while at the same time proposing a technique on top of Wikipedia hyperlink[3] structure to determine context of a tweet. Furthermore, we utilize the Wikipedia category graph structure of an entity to observe the amount of discussion related to an entity within the tweet[4].

The remainder of the paper is organized as follows. In Section 2 we provide an overview of related work. In Section 3 we describe Wikipedia with an emphasis on the features that are important to the entity name disambiguation task within online reputation management. In Section 4 we present a detailed description of the proposed methodology. In Section 5 we present experimental evaluations. Finally in Section 6, we provide some conclusions and discuss the implications of our findings from the point of view of usefulness of Wikipedia's hyperlink and category structure for entity name disambiguation in tweets.

---

[1] The task is organized as a CLEF evaluation task [1] where teams were given a set of entities and for each entity a set of tweets were provided, the challenge was to classify tweets as relevant or irrelevant with respect to the entity.

[2] We give more details about this in Section 3 and 4.

[3] Inlinks and outlinks within the Wikipedia articles.

[4] A tweet related to "Apple" may contain terms such as iOS, iPhone etc. which could easily be extracted from the Wikipedia category graph (i.e., subcategories) for Apple Inc. We will explain this in later sections.

2   Related Work

In this section we first provide an overview of some works that utilize Wikipedia for entity name disambiguation in long texts, and then we describe the works relevant to the entity name disambiguation task in tweets.

Bunesco and Pasca [4] proposed the utilization of Wikipedia for the entity name disambiguation task in texts by means of a similarity measure that compares the context of an entity occurrence to the Wikipedia categories of candidate entities. The technique proposed by Milne and Ian [9] employs supervised classification for mapping entity occurrences in a text to entities in Wikipedia via learned feature weights instead of directly employing the similarity function proposed by Bunesco and Pasca. [4]. They do so through a notion of semantic relatedness between candidate entities corresponding to the entity occurrence and the unambiguous mentions in the textual context. The relatedness values are derived from the overlap of incoming links in Wikipedia articles. Han and Zhao [7] also take into account the relatedness of common noun phrases in the context of an entity occurrence, matched against Wikipedia article titles. These approaches achieved very good results in experimental evaluations for long texts but they have been shown to perform poorly on tweets as described by Meij et al. [8].

In spite of the great significance of extracting commercially useful information from tweets, the amount of research dedicated to entity name disambiguation in tweets is very limited. Only two serious efforts have been undertaken which are by Ferragina and Scaiella [6] and Meij et al. [8]; both these approaches use Wikipedia for the task at hand. The TAGME system [6] uses the hyperlink structure of Wikipedia by exploiting the links between Wikipedia pages and the anchor texts of the links to those Wikipedia pages. Disambiguation is performed by application of a collective agreement function (i.e., a voting function) among all senses associated to anchors detected on the input texts and similar to the work of Meij et al [9], unambiguous anchors are utilized to boost the selection of these senses for the ambiguous anchors. Meij et al. [8] employ supervised machine learning techniques for refinement of a list of candidate Wikipedia concepts that are potentially relevant to a given tweet. The candidate ranking list is generated by matching n-grams in the tweet with anchor texts in Wikipedia articles, taking into account the hyperlink structure in Wikipedia to compute the most probable Wikipedia concept for each n-gram. Our approach differs from the above approaches in that we do not rely on the anchor text structure within Wikipedia, instead we directly employ inlinks and outlinks between Wikipedia articles along with the Wikipedia category graph. We show that our technique significantly outperforms the approach based on Meij et al.'s work within the CLEF 2013 RepLab filtering subtask [1].

3   Background

The proposed filtering algorithm makes use of the encyclopedic structure in Wikipedia; more specifically the knowledge encoded in Wikipedia's graph

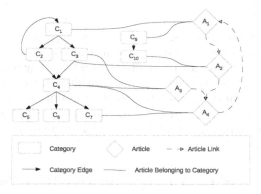

Fig. 1. Wikipedia Category Graph Structure along with Wikipedia Articles

structure is utilized for the classification of tweets as relevant or irrelevant with respect to a particular entity. Wikipedia categories are organized into a taxonomy-like[5] structure (see Figure 1). Each Wikipedia category can have an arbitrary number of subcategories as well as being mentioned inside an arbitrary number of supercategories (e.g., category C_4 in Figure 1 is a subcategory of C_2 and C_3, and a supercategory of C_5, C_6 and C_7.) Furthermore, in Wikipedia each article can belong to an arbitrary number of categories. As an example, in Figure 1, article A_1 belongs to categories C_1 and C_9, article A_2 belongs to categories C_3 and C_{10}, while article A_3 belongs to categories C_3 and C_4. In addition to links between Wikipedia categories and Wikipedia articles, there are also links between Wikipedia articles as the dotted lines in Figure 1 show (e.g., article A_1 outlinks to A_2 and has an inlink from A_4). The Wikipedia categories serve as a semantic tag for the articles to which they link [12]. Similarly, the inlinks and outlinks between Wikipedia articles are organized according to the semantics inside the articles' content (e.g., the article on "Apple Inc." has an inlink from the article on "Steve Jobs" while having an outlink the to article on "iPhone"). The technique we propose in this paper makes use of both the underlying semantics within the Wikipedia articles' hyperlink structure, and the category-article structure as we will explain in the next section. Finally, similar to Bunesco and Pasca [4] we utilize Wikipedia disambiguation pages corresponding to a Wikipedia article; an example is shown in Table 1 where corresponding to the entity "Apple" and "BMW" we show some disambiguation pages (senses) from Wikipedia. From here on, we use the terms disambiguation pages and senses to refer to the same concept.

Table 1. Disambiguation Pages (Senses) for Two Entities

Entity	Disambiguation Pages
Apple	Apple Corps, Apple Inc., Apple Bank, Apple (album), Apple (band), Apple (fruit) etc.
BMW	Bayerische Motoren Werke AG, BMW Open, BMW Championship (PGA Tour) etc.

---

[5] We say taxonomy-like because it is not strictly hierarchical due to the presence of cycles in the Wikipedia category graph.

4   Methodology

The proposed entity name disambiguation strategy employs a supervised classification model that utilizes features defined using the Wikipedia graph structure. The proposed approach involves a two-step method for entity name disambiguation. In the first step, we determine the phrases within a tweet using an approach similar to Meij et al. [8]. In the second step, we use the Wikipedia graph structure (i.e., both Wikipedia articles' hyperlink structure, and the Wikipedia category-article structure) to extract a set of features that enables us to perform the disambiguation task.

4.1   Feature Set Based on Wikipedia Articles' Hyperlinks

In this section we present our strategy to exploit the Wikipedia articles' hyperlink structure; first we discuss phrase extraction which is followed by a discussion on how we actually exploit the Wikipedia articles' hyperlink structure.

Phrase Extraction from Tweets. There are two kinds of phrases that we extract in this step. First is the entity phrase which represents the entity while the rest of the phrases are context phrases. As an example, consider the tweets in Table 2. For the first tweet, we extract all possible n-grams within the chunks "I prefer Samsung over HTC", "Apple", "Nokia", and "because it is economical and good". In this tweet, "Samsung" constitutes an entity phrase whereas other possible n-grams are considered as context phrases.

Context phrase extraction is performed by the generation of possible n-grams within phrase chunks of a tweet. We do not perform n-gram generation for the complete tweet but instead treat a tweet as a composition of phrase chunks with boundaries such as commas, semi-colons, sentence terminators etc. along with other tweet-specific markers such as @, RT etc. Similar to the technique proposed by Meij et al. [8] we then reduce candidate phrases extracted from a tweet to those that have a match in Wikipedia article titles[6]. The reduced set of phrases extracted from a tweet are referred to as *ContextPhrases*. In Table 2, considering the second tweet, we extract all possible n-grams within the chunks "Dear Ryanair", "I hate travelling with you", and "You suck." Note that we utilize the n-grams within tweets' phrase chunks for efficiency purposes in order to speed up the feature extraction process of the next step.

Table 2. Example Tweets to Illustrate Phrase Extraction

Entity	Tweet
Samsung	I prefer Samsung over HTC, Apple, Nokia because it is economical and good
Ryanair	Dear Ryanair, I hate travelling with you. You suck!!!

---

[6] We differ in that we do not apply supervised machine learning for reduction of candidate phrases.

Feature Extraction Using Wikipedia Articles' Hyperlinks. As described in the previous section, we extract an entity phrase and context phrases for each tweet which we now utilize in this section to generate features using the links between Wikipedia articles.

At the first level, we use the parent Wikipedia article for the entity under investigation[7] and we extract a set of parent categories that contain the entity name. For example, corresponding to entity "Toyota", the categories "Companies listed on the New York Stock Exchange", "Marine engine manufacturers", "Military equipment of Japan","Companies based in Nagoya" and "Toyota" occur as parent categories of which only "Toyota" is selected. We then extract sub-categories from the selected categories up to a depth count of two[8]; finally all articles belonging to these sub-categories are marked as being related to the entity under investigation and we refer these articles as $Articles_{related}$.

We then construct an information table of Wikipedia-based features using entity phrase, context phrases, and $Articles_{related}$ as follows:

- In order to perform entity disambiguation for the entity phrase, we extract from Wikipedia the disambiguation pages (senses) for an entity phrase and context phrases. Using these potential senses of the entity phrase (denoted as e_{s_i}) and each context phrase (denoted as c_{s_i}) we then define three collections or bags;
  - Wikipedia articles linking to e_{s_i} or any c_{s_i} referred to as *Inlinks*
  - Wikipedia articles linking from e_{s_i} or any c_{s_i} referred to as *Outlinks*
  - Wikipedia articles linking to/from e_{s_i} or any c_{s_i} referred to as *Inlinks+Outlinks*
- Using information of *Inlinks*, *Outlinks* and *Inlinks+Outlinks*, we derive the features shown in Table 3.

Table 3. Feature set for entity name disambiguation in tweets on top of Wikipedia Article Link Structure

Feature	Description
$Intersection_{duplication}$	No. of intersections between *Inlinks* for e_{s_i} and each c_{s_i} without removing duplicated articles
$NormalizedIntersection_{duplication}$	No. of intersections between *Inlinks* for e_{s_i} and each c_{s_i} without removing duplicated articles and normalized by total number of articles in the sets
$Intersection_{noduplication}$	No. of intersections between *Inlinks* for e_{s_i} and each c_{s_i} after removing duplicated articles
$NormalizedIntersection_{noduplication}$	No. of intersections between *Inlinks* for e_{s_i} and each c_{s_i} without removing duplicated articles and normalized by total number of articles in the sets
$Ratio_{inlink:outlink}$	Ratio between articles in *Inlinks* to articles in *Outlinks*

*Note that we calculate similarly for *Outlinks* and *Inlinks+Outlinks* for the first four features

---

[7] The parent Wikipedia article for each entity is given as part of the dataset for this task.

[8] This was chosen following empirical analysis; a depth of two was found sufficient to gather a representative set of categories while preventing too much drift.

– We illustrate the use of this entire feature set with the help of the example illustrated in Table 4 depicting a tweet with three context phrases $c1$, $c2$, and $c3$. Here, the entity phrase e has three Wikipedia senses e_{s_1}, e_{s_2}, and e_{s_3}. There are two senses corresponding to $c1$ (i.e, $c1_{s_1}$ and $c1_{s_2}$), three senses corresponding to $c2$ (i.e., $c2_{s_1}$, $c2_{s_2}$, and $c2_{s_3}$), and finally $c3$ which is unambiguous (i.e., has only one sense $c3_{s_1}$). We assume Table 4 to represent $Intersection_{noduplication}$ i.e., the third feature from Table 3 for $Inlinks$. Corresponding to each context phrase, the sense that maximizes $Intersection_{noduplication}\{Inlinks\}$ is chosen implying selection of $c1_{s_2}$ and $c2_{s_3}$ across e_{s_1}, $c1_{s_1}$ and $c2_{s_2}$ across e_{s_2} and finally, $c1_{s_1}$ and $c2_{s_3}$ across e_{s_3}; note that no reduction takes place for $c3$ on account of it having a single sense only. We show the reduction step in Table 5. The reduction is followed by averaging the numerical values of features (i.e., $Intersection_{noduplication}\{Inlinks\}$ in the considered example) for selected context phrase sense across each entity phrase sense implying a value of 294 across e_{s_1}, 318.33 across e_{s_2}, and 323.67 across e_{s_3}. As a final step, we select the entity phrase sense with the highest context phrase score and in the considered example e_{s_3} (with value 323.67) is selected and the value of this score is added as a feature for the entity name disambiguation task.

Furthermore, if the selected entity corresponding to the highest score value belongs to one of the articles that are related to the entity (i.e., articles in $Articles_{related}$ explained previously in this section), we add a Boolean feature marked True, and False otherwise. Hence, for each feature listed in Table 3 there are two associated features with one being a continuous variable (score) and the other being a discrete variable (Boolean value representing entity sense mapping). Note that this reduction of features is performed corresponding to each feature in Table 3 and for the purpose of the example above we only use $Intersection_{noduplication}\{Inlinks\}$; similarly it is done for all three $Inlinks$, $Outlinks$, and $Inlinks+Outlinks$. We also do such feature set construction separately for stemmed and non-stemmed versions of the tweets.

Table 4. Information Table Corresponding to $Intersection_{noduplication}\{Inlinks\}$

Entity	Context Phrase Senses					
Senses	c1		c2			c3
	$c1_{s_1}$	$c1_{s_2}$	$c2_{s_1}$	$c2_{s_2}$	$c2_{s_3}$	$c3_{s_1}$
e_{s_1}	150	230	400	415	532	120
e_{s_2}	180	147	350	375	280	400
e_{s_3}	234	115	83	127	237	500

Table 5. Reduction of Information Table Corresponding to $Intersection_{noduplication}\{Inlinks\}$

Entity	Context Phrase Senses		
Senses	c1	c2	c3
e_{s_1}	$c1_{s_2}$:230	$c2_{s_3}$:532	120
e_{s_2}	$c1_{s_1}$:180	$c2_{s_2}$:375	400
e_{s_3}	$c1_{s_1}$:234	$c2_{s_3}$:237	500

4.2 Relatedness Score Based on Wikipedia Category-Article Structure

In this section we present our strategy to exploit the Wikipedia category-article structure. First we extract possible n-grams from a tweet then we score relatedness of n-grams of a tweet with an entity. The relatedness scoring of n-grams utilizes both Wikipedia category and article structure to score the amount of content present inside the tweet related to an entity.

First, we fetch all the parent categories[9] and all sub-categories[10] to a depth count of two of an entity's Wikipedia article. We refer to these categories related to the entity under investigation as RC (i.e., it contains all related categories of an entity from depth count zero to two). Next, we retrieve the set of all articles within the Wikipedia category set RC (we refer this set as $Articles_{RC}$). Finally, all categories associated with these articles are retrieved which we refer to as WC; note[11] that RC is a subset of WC.

The extracted phrases from a tweet which are contained in $Articles_{RC}$ are called matched phrases. We use these matched phrases to calculate the relatedness score. The following summarizes important factors which contributes in calculating our relatedness score for a tweet using Wikipedia category-article structure.

- $Depth_{significance}$ denotes the significance of category depth at which a matched phrase occurs; the deeper the match occurs in the taxonomy the less its significance to the entity under consideration. This implies that the matched phrase in the parent category of the entity under investigation are more likely to be relevant to the entity than those at depth count of two.

$$Depth_{significance}(p) = \sum_{cat \in RC \cap p_{categories}} \frac{1}{depth_{cat} + 1}$$

- $Cat_{significance}$ denotes the significance of a matched phrase's categories corresponding to the entity under investigation. The more categories of matched phrase in RC, the higher the significance related to entity.

$$Cat_{significance}(p) = \frac{|RC \cap p_{categories}|}{|WC \cap p_{categories}|} * log(|RC \cap p_{categories}| + 1)$$

- $Phrase_{significance}$ is a combination of phrase word length and frequency of the phrase within the tweet. The greater the word length of a phrase, the more informative or important it becomes, likewise the more frequent the phrase is in the 140 characters of tweet, the more important the phrase is for a tweet.

$$Phrase_{significance}(p) = log(wordlen(p) + 1) * p_{frequency}$$

Finally the equation for calculating the relatedness score for a tweet based on Wikipedia category-article structure is:

$$= \sum_{p \in MatchedPhrases} Depth_{significance}(p) \times Cat_{significance}(p) \times Phrase_{significance}(p)$$

---

[9] These are basically the categories of an entity's Wikipedia article i.e., categories at the depth zero from the Wikipedia article of an entity .

[10] These are basically entity related categories at depth count of one and two.

[11] E.g., Wikipedia article "Steve Jobs" of "Apple Inc." contains a category "1955 births" which is not present either in parent nor in sub-categories of entity's Wikipedia article.

5   Experimental Evaluations

This section describes the experimental evaluations that we undertake to demonstrate the effectiveness of the proposed method. We perform supervised classification over the data set provided by the organizers of RepLab 2013 [1] in two different settings (i.e, two different models of training) with combinations of two different sets of features (i.e., those proposed in section 4.1 and 4.2).

5.1   Dataset

Tweets. The provided corpus is a multilingual collection of tweets (i.e., 20.3% Spanish tweets and 79.7% English tweets) for a set of 61 entities with entities belonging to four domains: automotive, banking, universities, and music. As a rough estimate for each entity at least 2200 tweets were collected with the first 700 constituting the training set, and the rest serving as the test set.

Wikipedia. The data for Wikipedia articles' hyperlinks, and Wikipedia category-article structure is obtained through a custom Wikipedia API that has pre-indexed Wikipedia data and hence, it is computationally fast[12]. The API is developed using the DBPedia [3] 2012 dumps. On account of dumps being based on 2012 Wikipedia dataset, one of the entities within the music domain, namely PSY, does not have any Wikipedia article[13] and hence, we perform experimental evaluations for 60 entities.

5.2   Experimental Setup

We split the training set into a development set (75% of the tweets in training set) and a validation set (25% of remaining tweets in training set). The split was performed through a randomly stratified approach per entity. The validation set is used to discover the best setting for each of the categories and entities.

The relatedness score feature based on Wikipedia category-article structure of section 4.2 (denoted as r) and these features combined with features of section 4.1 (this combination is denoted as hr) are used in conjunction with a random forest classifier by training over the four domains and the 60 entities. The motivation for distinguishing between domain based runs is to capture the notion of the vocabulary difference between various companies (e.g., terms like shares, loans are specifically related to banking-related companies). Furthermore, this vocabulary-based difference for entities belonging to different domains also occurs in Wikipedia articles' hyperlink and category-article structure which is natural for an encyclopaedic knowledge base. Following are the various settings which are used to find the optimal setting for each entity, and each domain.

---

[12] `http://bit.ly/1eMADG9`, we aim to release the API as an open source Wikipedia tool to facilitate other researchers.

[13] This singer became famous after the date of dumps and hence, his Wikipedia page did not exist before that.

- hr_{domain} uses features of both section 4.1 and 4.2 whilst training per domain i.e. combining all tweets related to a particular domain into one training and one test set
- hr_{entity} uses features of both section 4.1 and 4.2 whilst training per entity
- r_{domain} uses features of section 4.2 whilst training per domain
- r_{entity} uses features of section 4.2 whilst training per entity

Note that we do not use features of section 4.1 alone on account of previous experience during participation CLEF 2013 RepLab filtering task where results obtained with Wikipedia articles' hyperlinks features were promising but not among the best. Similarly, based on previous experience we choose not to train on a global model with all the training data at once.

5.3 Evaluation Measures

The measures used in the evaluation purposes are Reliability and Sensitivity, which are described in detail by Amigo et al. [2]. The property that makes them particularly suitable for the filtering problem is that they are strict with respect to standard measures, i.e., a high value according to Reliability and Sensitivity implies a high value in all standard measures.

5.4 Results

Tables 6-9 present the results of our experiments on the validation set. A separate table is used for each domain. Each table shows in boldface font the optimal setting that maximizes the F-measure of Reliability and Sensitivity. The results clearly demonstrate that for the automotives and universities domains based classification works best, and for the banking and music domains, entity based classification works best. We discuss this behavior in the next section. Table 10 shows the results for test dataset when keeping optimal settings learned from the validation dataset and finally, Table 11 shows our results' comparison to other participants of CLEF 2013 RepLab filtering task[14]. The system UvA UNED is the one that utilizes Wikipedia semantic linking features via an established system defined by Meij et al. [8], and our technique significantly outperforms it.

Table 6. Evaluation Results on Validation Set for Automotives

Setting	Reliability	Sensitivity	F-Measure
hr_{domain}	**0.64**	**0.58**	**0.61**
hr_{entity}	0.69	0.41	0.45
r_{domain}	0.58	0.52	0.55
r_{entity}	0.76	0.34	0.39

Table 7. Evaluation Results on Validation Set for Banking

Setting	Reliability	Sensitivity	F-Measure
hr_{domain}	0.38	0.51	0.38
hr_{entity}	**0.81**	**0.51**	**0.57**
r_{domain}	0.44	0.61	0.44
r_{entity}	0.77	0.51	0.55

---

[14] Note that our earlier participation was not amongst the top five systems but the extended features enable us to achieve second best results.

Table 8. Evaluation Results on Validation Set for University

Setting	Reliability	Sensitivity	F-Measure
hr_{domain}	**0.62**	**0.51**	**0.51**
hr_{entity}	0.63	0.38	0.42
r_{domain}	0.59	0.46	0.48
r_{entity}	0.65	0.34	0.39

Table 9. Evaluation Results on Validation Set for Music

Setting	Reliability	Sensitivity	F-Measure
hr_{domain}	0.46	0.28	0.31
hr_{entity}	0.84	0.29	0.38
r_{domain}	0.13	0.21	0.12
r_{entity}	**0.86**	**0.38**	**0.45**

Table 10. Evaluation Results on Test Set by Domain

Domain	Setting	Reliability	Sensitivity	F(R,S)
Automotives	hr_{domain}	0.54	0.47	0.47
Banking	hr_{entity}	0.75	0.58	0.49
University	hr_{domain}	0.71	0.44	0.49
Music	r_{entity}	0.83	0.34	0.39

Table 11. Performance Comparison with Other Systems

Team	Reliability	Sensitivity	F(R,S)
POPSTAR	0.73	0.45	0.49
OUR APPROACH	0.67	0.42	0.45
SZTE NLP	0.60	0.44	0.44
LIA	0.66	0.36	0.38
BASELINE	0.49	0.32	0.33
UvA UNED	0.68	0.22	0.21

6   Discussion and Future Work

The experimental evaluations establish Wikipedia's strength as a significant encyclopaedic resource for the task of entity name disambiguation in tweets. We now explain the reasons behind the difference in performance for domain based classification and entity based classification over the given dataset. The performance for banking and music is improved when using entity based classification on account of the different nature of each included entity (e.g., the banking dataset includes banks from England, Scotland, Spain, and United States; similarly the music dataset has a diverse range of singers from "Lady Gaga" to "Beatles"). On the other hand, the automotives and university dataset includes entities that are more homogeneous (e.g., the automotives dataset includes all entities that are famous automobile brands and deal in more or less the same products; similarly the university dataset includes all universities that are top in rankings with most being from United States). The Wikipedia articles' hyperlink features and category-article relatedness score captures the dataset differences in a powerful manner due to its organization as an encyclopaedic knowledge base.

The relatedness score defined using Wikipedia category-article structure introduces a powerful semantic notion of linking n-grams in a tweet with the information relevant to an entity under discussion as shown by evaluations on music domain. On the other hand, features extracted from articles' hyperlinks albeit powerful introduce some amount of noise where chatter and gossip about musicians leads to noisy inlinks and outlinks.

Other teams such as POPSTAR and LIA utilized text similarity based features while LIA also used Wikipedia based similarity measures [1]. As future work, we aim to combine our Wikipedia based features with text based techniques to further improve the performance of our proposed entity name disambiguation for tweets.

References

1. Amigó, E., Carrillo de Albornoz, J., Chugur, I., Corujo, A., Gonzalo, J., Martín, T., Meij, E., de Rijke, M., Spina, D.: Overview of replab 2013: Evaluating online reputation monitoring systems. In: Forner, P., Müller, H., Paredes, R., Rosso, P., Stein, B. (eds.) CLEF 2013. LNCS, vol. 8138, pp. 333–352. Springer, Heidelberg (2013)
2. Amigó, E., Gonzalo, J., Verdejo, F.: A General Evaluation Measure for Document Organization Tasks. In: Proceedings SIGIR (July 2013)
3. Bizer, C., Lehmann, J., Kobilarov, G., Auer, S., Becker, C., Cyganiak, R., Hellmann, S.: Dbpedia - a crystallization point for the web of data. Web Semant 7(3), 154–165 (2009)
4. Bunescu, R.C., Pasca, M.: Using encyclopedic knowledge for named entity disambiguation. In: EACL, vol. 6, pp. 9–16 (2006)
5. Dellarocas, C., Awad, N.F., Zhang, X.M.: Exploring the value of online reviews to organizations: Implications for revenue forecasting and planning. In: Management Science, pp. 1407–1424 (2003)
6. Ferragina, P., Scaiella, U.: Tagme: on-the-fly annotation of short text fragments (by wikipedia entities). In: CIKM 2010, pp. 1625–1628. ACM, New York (2010)
7. Han, X., Zhao, J.: Named entity disambiguation by leveraging wikipedia semantic knowledge. In: CIKM 2009, pp. 215–224. ACM, New York (2009)
8. Meij, E., Weerkamp, W., de Rijke, M.: Adding semantics to microblog posts. In: WSDM 2012, pp. 563–572. ACM, New York (2012)
9. Milne, D., Witten, I.H.: Learning to link with wikipedia. In: CIKM 2008, pp. 509–518. ACM (2008)
10. Peetz, M.-H., Spina, D., Gonzalo, J., de Rijke, M.: Towards an active learning system for company name disambiguation in microblog streams. In: CLEF (Online Working Notes/Labs/Workshop) (2013)
11. Qureshi, M.A., Younus, A., Abril, D., O'Riordan, C., Pasi, G.: Cirg irdisco at replab2013 filtering task: Use of wikipedia's graph structure for entity name disambiguation in tweets. In: CLEF (Online Working Notes/Labs/Workshop) (2013)
12. Zesch, T., Gurevych, I.: Analysis of the Wikipedia Category Graph for NLP Applications. In: Proceedings of the TextGraphs-2 Workshop, NAACL-HLT (2007)

From Treebank Conversion
to Automatic Dependency Parsing for Vietnamese

Dat Quoc Nguyen[1], Dai Quoc Nguyen[1], Son Bao Pham[1], Phuong-Thai Nguyen[1],
and Minh Le Nguyen[2]

[1] Faculty of Information Technology,
University of Engineering and Technology,
Vietnam National University, Hanoi
{datnq,dainq,sonpb,thainp}@vnu.edu.vn
[2] School of Information Science,
Japan Advanced Institute of Science and Technology
nguyenml@jaist.ac.jp

Abstract. This paper presents a new conversion method to automatically transform a constituent-based Vietnamese Treebank into dependency trees. On a dependency Treebank created according to our new approach, we examine two state-of-the-art dependency parsers: the MSTParser and the MaltParser. Experiments show that the MSTParser outperforms the MaltParser. To the best of our knowledge, we report the highest performances published to date in the task of dependency parsing for Vietnamese. Particularly, on gold standard POS tags, we get an unlabeled attachment score of 79.08% and a labeled attachment score of 71.66%.

1 Introduction

Dependency parsing is one of the major research topics in natural language processing (NLP) as dependency-based syntactic representations are useful for many NLP applications such as machine translation and information extraction [1]. This research field was boosted by the successes of the CoNLL shared tasks on multilingual dependency parsing [2,3] where raising current state-of-the-art approaches based on supervised data-driven machine learning. McDonald and Nivre [4] has determined two major categories of data-driven dependency parsing: graph-based approaches [5,6,7,8] and transition-based ones [9,10,11]. In addition, there are hybrid methods [12,13] to combine the graph-based and transition-based approaches. However, those methods require such large training corpora as dependency Treebanks which are very expensive: taking a lot of time and human effort to manually annotate the corpora.

In many languages like Vietnamese, there are no manually labeled dependency Treebanks available. Since constituent structure-based Treebanks, for instances the English Penn Treebank [14] and the Vietnamese Treebank [15], are dominant resources to develop natural language parsers, constituent-to-dependency conversion approaches must be applied to generate larger amounts of annotated dependency structure-based corpora. Johansson and Nugues [16] proposed an extended conversion procedure for English to overcome drawbacks of previous methods [9,17] on new versions of the English Penn

E. Métais, M. Roche, and M. Teisseire (Eds.): NLDB 2014, LNCS 8455, pp. 196–207, 2014.

Treebank. The Treebank transformation procedures in such other languages as German, French, Spanish, Bulgarian, Chinese and Korean can be correspondingly found in [18], [19], [20], [21], [22] and [23].

Turning to Vietnamese, there are only two works [24,25] on dependency parsing. Hong et al. [24] described an approach for Vietnamese dependency parsing based on Lexicalized Tree-Adjoining Grammars (LTAG). A LTAG parser was trained on set of 8367 constituent trees in the Vietnamese Treebank. Then dependency relations were extracted from derivation trees returned by LTAG parsing. Evaluating on 441 sentences of 30-words length or less, the Hong et al. [24]'s method obtained an unlabeled attachment score[1] of 73.21% on automatically assigned part-of-speech (POS) tags.

Thi et al. [25] presented a constituent-to-dependency transformation method for Vietnamese. The method applied head-percolation rules constructed for Vietnamese as exhibited in [26] to find the head of each constituent phrase. However, it is not clearly how dependency labels are inferred since Thi et al. [25] just outlined that there was a function namely GetDependentLabel exploited to label dependency relations. On a Stanford format-based dependency Treebank of 10k sentences converted from the Vietnamese Treebank [15] according to their own procedure, Thi et al. [25] showed good experimental results on 10-fold cross validation evaluation scheme. They earned 73.03% and 66.35% computed for the unlabeled and labeled attachment scores given by the transition-based MaltParser toolkit [11] using gold standard POS tags.

The difference between our work and the two previous works on Vietnamese dependency parsing is that we make a better use of existing information in the Vietnamese Treebank. The previous works performed shallow processes as they do not take grammatical function tags into account as well as do not handle cases of coordination and empty category mappings. To sum up, the contributions of our study are:

• We propose a new constituent-to-dependency conversion method to automatically create a dependency Treebank from the input constituent-based Vietnamese Treebank. Specifically, in addition to modifications of head-percolation rules, we bring clear heuristics to label dependency relations employing grammatical function tags and other existing information in the Vietnamese Treebank. We also solve the cases of coordination and empty categories.

• We provide[2] a Vietnamese dependency Treebank namely VnDT containing 10200 sentences. The VnDT Treebank is formatted following 10-column data format as proposed by the CoNLL shared tasks on multilingual dependency parsing [2,3]. It is because most state-of-the-art dependency parsers such as the graph-based MSTParser [5] and the MaltParser refer to the CoNLL 10-column format as an input standard.

• We achieve highest performances published to date in the task of Vietnamese dependency parsing by examining the MSTParser and the MaltParser on our VnDT Treebank. In particular on the evaluation scheme of 10-fold cross validation, we gain the unlabeled attachment score (UAS) of 79.08% and the labeled attachment score (LAS) of 71.66% on gold standard POS tags. Besides, the highest parsing scores are 76.21% for UAS and 66.95% for LAS in terms of automatically assigned POS tags.

---

[1] Un-analyzable sentences and punctuations are not taken into account.

[2] Our Vietnamese dependency parsing resources including preprocessing tools, pre-trained models and the VnDT Treebank are available at http://vndp.sourceforge.net/

2   Our New Treebank Conversion Approach for Vietnamese

This section is to introduce our new procedure to automatically convert constituent trees to dependency trees for Vietnamese. We provide information about the Vietnamese constituent trees in section 2.1 and technique to transform the constituent trees to unlabeled dependency trees in section 2.2. We then describe how our method labels dependency links in the use of function tags in section 2.3 and heuristics to infer suitable labels in section 2.4. The processes of solving cases of coordination and empty category mappings are detailed in sections 2.5 and 2.6, respectively.

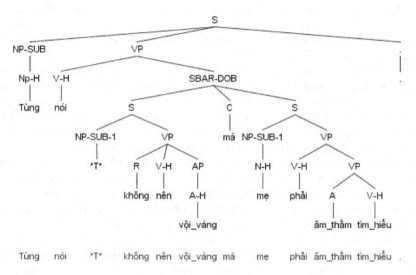

Fig. 1. A tree in the Vietnamese Treebank *"Tùng nói *T* không nên vội_vàng mà mẹ phải âm_thầm tìm_hiểu ."* (*Tung suggests that *T* should not be hastily but mother must be silent to find out .*)

2.1   Constituent Trees in the Vietnamese Treebank

The Vietnamese Treebank [15] has been produced as a part of the national project VLSP[3]. It contains about 10200 constituent trees (220k words) formatted similarly as those trees in the Penn Treebank [14]. In figure 1 illustrating a constituent tree of a Vietnamese sentence, the tree includes Vietnamese words at leaf nodes[4], and POS tag-level and phrase-level nodes as explained in table 1. Table 1 also gives information about grammatical function tags associated to POS and phrase-level tags at non-leaf nodes. These tags will be used in our transformation approach to label dependency relations.

---

[3] http://vlsp.vietlp.org:8080/

[4] Vietnamese is a monosyllabic language; hence, a word may consist of more than one token. Tokens in a word can be distinguished by underline character _. The *T* at a leaf node in figure 1 means an empty category.

Table 1. Part-Of-Speech (POS) tags, phrase-level and function tags existing in the Vietnamese Treebank. In addition to LBKT and RBKT (left/right bracket) POS tags, there are such different punctuation types as .,;:-"/ and ...

POS tags			
N Noun	Nb Borrowed noun	R Adjunct	I Exclamation
Np Proper noun	V Verb	L Determiner	T Particle
Nc Classifier noun	Vb Borrowed verb	M Quantity	Y Abbreviation
Nu Unit noun	A Adjective	E Preposition	S Affix
Ny Abbreviated noun	P Pronoun	C Conjunction	X Un-definition/Other

Phrase-level tags		Function tags			
S	Sentence	UCP	Unlike coordinated phrase	H	Head	TMP	Temporal
SQ	Question	YP	Abbreviation phrase	SUB	Subject	LOC	Location
SBAR	Subordinate clause	WHNP	Wh-noun phrase	DOB	Direct object	DIR	Direction
NP	Noun phrase	WHAP	Wh-adjective phrase	IOB	Indirect object	MNR	Manner
VP	Verb phrase	WHRP	Wh-adjunct phrase	TPC	Topicalization	PRP	Purpose
AP	Adjective phrase	WHPP	Wh-prepositional phrase	PRD	Predicate	CND	Condition
RP	Adjunct phrase	WHVP	Wh-verb phrase	EXT	Extent	CNC	Concession
PP	Prepositional phrase	WHXP	Wh-undefined phrase	VOC	Vocatives	ADV	Adverbial
QP	Quantity phrase	XP	Un-definition/other phrase				
MDP	Modal phrase						

2.2   Head-Percolation Rules

Finding the head of each phrase is an essential task in order to generate dependency links. Similar to a common manner to find the head in a phrase structure [9,17], our method is based on a classical technique of exploiting head-percolation rules (head rules). Following [25], we employ the head rules built for Vietnamese as shown in [26].

There are around 2200 unindexed empty categories such as *(NP-SUB *T*)*, *(NP-SUB *E*)* and *(V-H *E*)* appearing in the Vietnamese Treebank. For those unindexed phrases, it is unable to retrieve the corresponding phrases in an empty category mapping process. It leaded to a removal of those phrases from the Vietnamese Treebank. Therefore, we made minor changes on some existing head rules to adapt to the modified Vietnamese Treebank. For example, we changed the rule for VP by adding SBAR, R, RP and PP.

In table 2 presenting the used head rules: the first column denotes the phrase types, the second one indicates a search direction (l or r expressing a looking for leftmost or rightmost constituent respectively), and the third is a left-driven priority list of phrase-level and POS tags to look for (.* meaning any tag while ; be a delimiter between tags). For example, to determine the head of a S phrase node, we look from left to right for any its child node associated to H function tag. If no child node H is found, we find for any child node with a S tag, and so on.

Using the head rules, it is straightforward to create unlabeled dependency trees from constituent trees: (i) marking the head of each phrase structure utilizing its head rule, and (ii) making dependents on the head for other child nodes in the phrase. For instance, a dependency tree consisting of unlabeled relation links will be returned as demonstrated in figure 2 for the input constituent tree in figure 1.

Table 2. Head-percolation rules for Vietnamese

S	l	*-H;S;VP;AP;NP;.*
SBAR	l	*-H;SBAR;S;VP;AP;NP;.*
SQ	l	*-H;SQ;VP;AP;NP;WHPP;.*
NP	l	*-H;NP;Nc;Nu;Np;N;P;VP;.*
VP	l	*-H;VP;V;A;AP;N;NP;SBAR;S;R;RP;PP;.*
AP	l	*-H;AP;A;N;S;.*
RP	r	*-H;RP;R;T;NP;.*
PP	l	*-H;PP;E;VP;SBAR;AP;QP;.*
QP	l	*-H;QP;M;.*
XP	l	*-H;XP;X;.*
YP	l	*-H;YP;Y;.*
MDP	l	*-H;MDP;T;I;A;P;R;X;.*
WHNP	l	*-H;WHNP;NP;Nc;Nu;Np;N;P;.*
WHAP	l	*-H;WHAP;A;N;V;P;X;.*
WHRP	l	*-H;WHRP;P;E;T;X;.*
WHPP	l	*-H;WHPP;E;P;X;.*
WHXP	l	*-H;XP;X;.*
WHVP	l	*-H;V;.*
UCP	l	*-H;.*

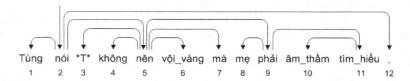

Fig. 2. An unlabeled dependency tree converted from the tree in figure 1 in using the head rules for Vietnamese

2.3 Grammatical Dependency Labels

We exploit all grammatical function tags as dependency labels excluding the tag *H* as *H* is employed in head-percolation rules to determine the head of every phrase. Some of the function tags may be combined[5] together like *TMP-TPC*. However, as pointed out by Choi and Palmer [27], most statistical dependency parsers do not often detect joined tags precisely. Hence, our conversion procedure keeps only the first function one in the pair of combined tags. Taking *TMP-TPC* as an example, the tag *TMP* will be selected as the dependency label instead of the joined tag *TMP-TPC*.

2.4 Inferred Labels

Most of dependency links (arcs) in converted trees have no label. In order to label those links, we use heuristic rules as detailed in our algorithm 1.

---

[5] There are about 200 joined-tags pairs in the Vietnamese Treebank.

Algorithm 1. Rules to label dependency arcs

Data: Let c be a word while C is the highest node for which c is the head of. And P is the parent of C with the word p be the head of P.

Result: A dependency label for the arc $c \longleftarrow p$

if *C is the root node* **then return** ROOT

else if *C is X, XP or WHXP* **then**

    if *C has a function tag of non-H* **then return** X + the tag; // There are XADV,
    XLOC, XMDP, XTMP, XPRD and XMNR.

    else return X

else if *C has a function tag of non-H* **then** // Section 2.3: 15 grammatical
function tags as dependency labels

    return the function tag;

else if *c is a determiner* **then return** DET

else if *c is a punctuation* **then return** PUNCT

else if *P is VP, and C is E, R, RP or WHRP* **then return** ADV

else if *P is VP or WHVP* **then return** VMOD; // Verb modifier

else if *P is AP or WHAP* **then return** AMOD; // Adjective modifier

else if *P is NP or WHNP* **then return** NMOD; // Noun modifier

else if *P is PP, and C is NP* **then return** POB; // Object of a preposition

else if *P is PP or WHPP* **then return** PMOD; // Prepositional modifier

else return *DEP*; // Default label

Figure 3 displays a dependency tree with labels for which it is transformed from the constituent tree in figure 1 in the use of the head rules and algorithm 1.

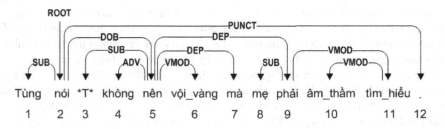

Fig. 3. A labeled dependency tree transformed from the tree in figure 1 in utilizing the head rules and the algorithm 1

2.5   Coordination

Our new procedure refers to a phrase as a coordinated one[6] if: (i) the phrase contains at least a child labeled with a C tag (i.e. conjunction), and (ii) the heads of both left and the right conjuncts (i.e. the left and right siblings of the C node) have the same

---

[6] The commas and semicolons are considered as separators within a coordinated phrase. Due to the Vietnamese Treebank annotation guideline: the UCP phrase always has at least a C-tag child. There are just 20 UCP-tagged phrases in the Vietnamese Treebank.

POS type. For instance in figure 1, we have the node $SBAR$ be a coordination as it has child node C corresponding to the word "mà$_{but}$", and the heads of two left and right conjuncts S are verbs "nên$_{should}$" and "phải$_{must}$".

There are several ways to represent coordinations in dependency structure [28,29]. Because we aim to generate the corpus of dependency trees in the CoNLL 10-column format, we follow the CoNLL dependency approach [2,3] to use dependency labels COORD and CONJ. Our method treats each preceding conjunct or conjunction to be the head of its following conjunct or conjunction. Figure 4 shows a dependency tree with a coordination example associated to the tree in figure 1.

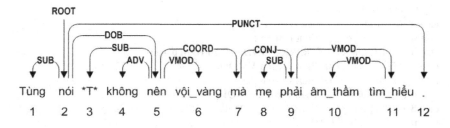

Fig. 4. A coordination example

2.6 Empty Categories

There are two types of empty categories in the Vietnamese Treebank including $*T*$ (trace) and $*E*$ (expletive). Because all $*E*$ phrases and some of $*T*$ phrases have no associated indexes to map, thus, we removed those ones as mentioned in section 2.2. Turning to the remaining $*T*$ indexed empty categories, our approach to map those $*T*$ phrases is similar to the conversion method for English as described in [16]: relinking the heads of the phrases which are referred to by the corresponding indexes.

For example, from the trees in figures 1 and 4, we relink the head "mẹ$_{mother}$" (#8) of the phrase (NP-SUB-1 (N-H mẹ)) to be a dependent of the word "nên$_{should}$" (#5) which the $*T*$ (#3) depends on. The dependency label for the $*T*$ (#3) will be the dependency label for the word "mẹ$_{mother}$" (#8) in this empty category mapping process. The final

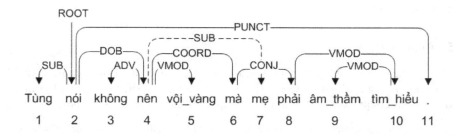

Fig. 5. The final output dependency tree

output dependency tree converted from the constituent tree in figure 1 according to our new transformation method is illustrated in figure 5 with the corresponding CoNLL 10-column format displayed in table 3. The relinking process creates some *non-projective* dependency trees (consisting of crossing links). Figure 5 presents an example of a non-projective tree.

Table 3. CoNLL format associated to the tree in figure 5

1	Tùng	_ N	Np	_	2	SUB	_ _
2	nói	_ V	V	_	0	ROOT	_ _
3	không	_ R	R	_	4	ADV	_ _
4	nên	_ V	V	_	2	DOB	_ _
5	vội_vàng	_ A	A	_	4	VMOD	_ _
6	mà	_ C	C	_	4	COORD	_ _
7	mẹ	_ N	N	_	4	SUB	_ _
8	phải	_ V	V	_	6	CONJ	_ _
9	âm_thầm	_ A	A	_	10	VMOD	_ _
10	tìm_hiểu	_ V	V	_	8	VMOD	_ _
11	.	_	.	_	2	PUNCT	_ _

3   Experiments on Dependency Parsing for Vietnamese

3.1   Experimental Setup

Data Set. We conducted experiments of Vietnamese dependency parsing on our VnDT Treebank consisting of 10200 trees (219k words). The VnDT Treebank is automatically converted from the input Vietnamese Treebank [15] based on our new conversion approach. The VnDT schema contains 33 dependency labels as mentioned in the previous section. Table 4 shows the distributions of the labels in the VnDT Treebank. The proportion of non-projective trees in VnDT is 4.49% while it is 80% accounted for the percentage of sentences of 30-words length or less.

Table 4. Distributions of dependency labels (in %): $X.^*$ means any dependency label starting with X while $.^*OB$ denotes any label ending by OB including DOB, IOB, and POB. *O.F.Tags* refers to other grammatical function tags as dependency labels.

ROOT	4.66	DET	6.19
VMOD	14.82	PUNCT	13.95
NMOD	19.01	DEP	3.13
AMOD	2.35	X.*	0.28
PMOD	0.24	.*OB	11.89
COORD	1.88	ADV	5.93
CONJ	1.86	SUB	6.78
O.F.Tags	7.03		

Evaluation Scheme. Results are evaluated on 10-fold cross validation scheme. We randomly separate the VnDT Treebank into 10 folds, giving one fold size of 1020 sentences. This evaluation procedure is repeated 10 times where each fold is used as the test set, and 9 remaining folds are merged as the training set. All our experimental results are reported as the average performances over the test folds.

Dependency Parsing Toolkits. We trained and evaluated two state-of-the-art dependency parsers: the graph-based MSTParser[7] [5,6] and the transition-based MaltParser [11]. As the MaltParser only produces projective dependency trees, we utilized the MaltOptimizer [30] - an optimization tool designed for the MaltParser to provide suitable feature model and parameters, and to select best parsing algorithms including nonprojective ones.

Metrics. Accuracy metrics[8] [3] are the unlabeled attachment score (UAS) and the labeled attachment score (LAS). UAS: The percentage of words that are correctly assigned the head (or no head if the word is a root). LAS: The percentage of words that are correctly assigned the head and dependency label (or no head if the word is a root).

3.2 Accuracy Results

Accuracies on Gold Standard POS Tags. Table 5 gives accuracies gained by the two parsers on gold standard POS tags, where the results computed on the set of short sentences (30-words length or less) and on the remaining longer sentences are also provided. It is clearly that the MSTParser surpasses the MaltParser. On UAS scores, the MSTParser obtains an accuracy of 79.08% which is 1.71% higher than the performance at 77.37% produced by the MaltParser. For LAS scores, the MSTParser does better with the result of 71.66% than the MaltParser returning a score of 70.49%.

Table 5. Accuracy results (%) on gold standard POS tags

	MST		Malt	
Length	UAS	LAS	UAS	LAS
<= 30 words	80.89	73.48	79.28	72.38
> 30 words	76.19	68.74	74.31	67.47
All	**79.08**	**71.66**	*77.37*	*70.49*

Because we have different set of dependency labels in comparison to the Thi et al. [25]'s dependency Treebank, it is not suitable in order to directly compare our method with the Thi et al. [25]'s one. However, on the same 10-fold cross validation scheme with the similar sizes of dependency corpora which are both transformed from the same Vietnamese Treebank, we reach higher performances of 4%+ improvements given by the MaltParser for which the Thi et al. [25]'s approach achieved the UAS of 73.03% and the LAS of 66.35%.

---

[7] The MSTParser is used with default parameter settings associated to "decode-type" of "nonproj".

[8] Accuracy results are calculated without scoring on punctuations.

Accuracies on Automatically POS Tagging. We also carried out the experiments on automatically assigned POS tags. We adapted the RDRPOSTagger toolkit[9] [31,32] to perform Vietnamese POS tagging with an accuracy result of 94.61% on the set of raw word-segmented sentences extracted from the VnDT Treebank.

The UAS and LAS results are presented in table 6. The highest scores are returned by the MSTParser with the UAS score of 76.21% and the LAS score of 66.95%.

Table 6. Accuracies (%) on automatically assigned POS tags

	MST		Malt	
Length	UAS	LAS	UAS	LAS
<= 30 words	77.85	68.60	76.30	67.52
> 30 words	73.59	64.31	71.66	62.97
All	**76.21**	**66.95**	*74.52*	*65.77*

Turning to the short sentences, we obtain the greatest UAS accuracy of 77.85% which is 4.64% higher than the UAS result of 73.21% reported by the Hong et al. [24]'s method evaluated on the 441 sentences of 30-words length or less. Though it is not on the same evaluation scheme, we show very promising results in the task of Vietnamese dependency parsing.

4   Conclusion

In this paper, we describe a new constituent-to-dependency conversion approach to automatically transform the Vietnamese Treebank [15] to dependency trees. Our procedure brings a better use of existing information in the Vietnamese Treebank.

We provide our Vietnamese dependency Treebank VnDT of 10200 sentences formatted following the CoNLL 10-column standard, and examine two state-of-the-art parsers the MSTParser and the MaltParser on the VnDT Treebank. Experiments point out that ' the MSTParser performs better than the MaltParser in Vietnamese dependency parsing task. To the best of our knowledge, we earn highest accuracy results published to date. For gold standard POS tags, we get the UAS score of 79.08% and the LAS score of 71.66%. On automatically assigned POS labels, the scores are 76.21% and 66.95% for the UAS and the LAS, respectively.

Acknowledgment. This work is partially supported by the Research Grant from Vietnam National University, Hanoi No. QG.14.04.

References

1. Kübler, S., McDonald, R., Nivre, J.: Dependency Parsing. Synthesis Lectures on Human Language Technologies. Morgan & Claypool Publishers (2009)

---

[9] http://rdrpostagger.sourceforge.net/

2. Buchholz, S., Marsi, E.: CoNLL-X shared task on multilingual dependency parsing. In: Proceedings of the Tenth Conference on Computational Natural Language Learning, CoNLL-X, pp. 149–164 (2006)
3. Nivre, J., Hall, J., Kübler, S., McDonald, R., Nilsson, J., Riedel, S., Yuret, D.: The CoNLL 2007 Shared Task on Dependency Parsing. In: Proceedings of the CoNLL Shared Task Session of EMNLP-CoNLL 2007, pp. 915–932 (2007)
4. McDonald, R., Nivre, J.: Characterizing the Errors of Data-Driven Dependency Parsing Models. In: Proceedings of the 2007 Joint Conference on Empirical Methods in Natural Language Processing and Computational Natural Language Learning (EMNLP-CoNLL), pp. 122–131 (June 2007)
5. McDonald, R., Pereira, F., Ribarov, K., Hajič, J.: Non-projective Dependency Parsing Using Spanning Tree Algorithms. In: Proceedings of the Conference on Human Language Technology and Empirical Methods in Natural Language Processing, HLT 2005, pp. 523–530 (2005)
6. McDonald, R., Lerman, K., Pereira, F.: Multilingual Dependency Analysis with a Two-stage Discriminative Parser. In: Proceedings of the Tenth Conference on Computational Natural Language Learning, CoNLL-X 2006, pp. 216–220 (2006)
7. Nakagawa, T.: Multilingual Dependency Parsing Using Global Features. In: Proceedings of the CoNLL Shared Task Session of EMNLP-CoNLL 2007, pp. 952–956 (2007)
8. Koo, T., Collins, M.: Efficient Third-order Dependency Parsers. In: Proceedings of the 48th Annual Meeting of the Association for Computational Linguistics, ACL 2010, pp. 1–11 (2010)
9. Yamada, H., Matsumoto, Y.: Statistical dependency analysis with support vector machines. In: Proceedings of the 8th International Workshop of Parsing Technologies, IWPT 2003 (2003)
10. Nilsson, J., Nivre, J., Hall, J.: Graph Transformations in Data-Driven Dependency Parsing. In: Proceedings of the 21st International Conference on Computational Linguistics and 44th Annual Meeting of the Association for Computational Linguistics, pp. 257–264 (July 2006)
11. Nivre, J., Hall, J., Nilsson, J., Chanev, A., Eryigit, G., Kübler, S., Marinov, S., Marsi, E.: MaltParser: A language-independent system for data-driven dependency parsing. Natural Language Engineering 13, 1 (2007)
12. Nivre, J., McDonald, R.: Integrating Graph-Based and Transition-Based Dependency Parsers. In: Proceedings of ACL 2008, pp. 950–958. HLT (June 2008)
13. Zhang, Y., Clark, S.: A Tale of Two Parsers: Investigating and Combining Graph-based and Transition-based Dependency Parsing. In: Proceedings of the 2008 Conference on Empirical Methods in Natural Language Processing, Honolulu, Hawaii, pp. 562–571 (October 2008)
14. Marcus, M.P., Marcinkiewicz, M.A., Santorini, B.: Building a large annotated corpus of English: the penn treebank. Comput. Linguist. 19(2), 313–330 (1993)
15. Nguyen, P.T., Vu, X.L., Nguyen, T.M.H., Nguyen, V.H., Le, H.P.: Building a Large Syntactically-Annotated Corpus of Vietnamese. In: Proceedings of the Third Linguistic Annotation Workshop, pp. 182–185 (August 2009)
16. Johansson, R., Nugues, P.: Extended Constituent-to-dependency Conversion for English. In: Proceedings of 16th Nordic Conference of Computational Linguistics, NODALIDA 2007, Tartu, Estonia, pp. 105–112 (2007)
17. Collins, M.: Three generative, lexicalised models for statistical parsing. In: Proceedings of the 35th Annual Meeting of the Association for Computational Linguistics, ACL 1997, pp. 16–23 (1997)
18. Seeker, W., Kuhn, J.: Making Ellipses Explicit in Dependency Conversion for a German Treebank. In: Proceedings of the 8th International Conference on Language Resources and Evaluation, LREC 2012, pp. 3132–3139 (2012)

19. Candito, M., Crabbé, B., Denis, P.: Statistical French dependency parsing: treebank conversion and first results. In: Proceedings of the 7th International Conference on Language Resources and Evaluation, LREC 2010 (2010)
20. Gelbukh, A., Calvo, H., Torres, S.: Transforming a constituency treebank into a dependency treebank. In: Proceedings of XXI Conference of the Spanish Society for Natural Language Processing, SEPLN 2005, vol. 35, pp. 145–152 (2005)
21. Marinov, S., Nivre, J.: A data-driven dependency parser for Bulgarian. In: Proceedings of the Fourth Workshop on Treebanks and Linguistic Theories (2005)
22. Ma, X., Zhang, X., Zhao, H., Lu, B.L.: Dependency Parser for Chinese Constituent Parsing. In: Joint Conference on Chinese Language Processing, pp. 1–6 (2010)
23. Choi, J.D., Palmer, M.: Statistical dependency parsing in Korean: from corpus generation to automatic parsing. In: Proceedings of the Second Workshop on Statistical Parsing of Morphologically Rich Languages, pp. 1–11 (2011)
24. Hong, P.L., Nguyen, T.M.H., Roussanaly, A.: Vietnamese Parsing with an Automatically Extracted Tree-Adjoining Grammar. In: Proceedings of the 9th IEEE RIVF International Conference on Computing & Communication Technologies, Research, Innovation, and Vision for the Future, pp. 1–6. IEEE (February 2012)
25. Thi, L.N., My, L.H., Viet, H.N., Minh, H.N.T., Hong, P.L.: Building a Treebank for Vietnamese Dependency Parsing. In: Proceedings of the 10th IEEE RIVF International Conference on Computing & Communication Technologies, Research, Innovation, and Vision for the Future (2013)
26. Le-Hong, P., Nguyen, T.M.H., Nguyen, P.T., Roussanaly, A.: Automated extraction of tree adjoining grammars from a treebank for Vietnamese. In: Proceedings of The Tenth International Workshop on Tree Adjoining Grammars and Related Formalisms (2010)
27. Choi, J.D., Palmer, M.: Robust constituent-to-dependency conversion for English. In: Proceedings of 9th Treebanks and Linguistic Theories Workshop, pp. 55–66 (2010)
28. de Marneffe, M.C., Manning, C.D.: The Stanford typed dependencies representation. In: Proceedings of the Coling 2008 workshop on Cross-Framework and Cross-Domain Parser Evaluation. Number, pp. 1–8 (2008)
29. Čmejrek, M., Cu\vr\'in, J., Havelka, J.: Prague Czech-English Dependency Treebank: Any Hopes for a Common Annotation Scheme? In: HLT-NAACL 2004 Workshop: Frontiers in Corpus Annotation, pp. 47–54 (May 2004)
30. Ballesteros, M., Nivre, J.: MaltOptimizer: A System for MaltParser Optimization. In: Proceedings of the 8th International Conference on Language Resources and Evaluation, LREC 2012, vol. (2006), pp. 2757–2763 (2012)
31. Nguyen, D.Q., Nguyen, D.Q., Pham, S.B., Pham, D.D.: Ripple Down Rules for Part-of-Speech Tagging. In: Proceedings of the 12th International Conference on Computational Linguistics and Intelligent Text Processing, CICLing 2011, vol. Part I, pp. 190–201 (2011)
32. Nguyen, D.Q., Nguyen, D.Q., Pham, D.D., Pham, S.B.: RDRPOSTagger: A Ripple Down Rules-based Part-Of-Speech Tagger. In: Proc. of the Demonstrations at the 14th Conference of the European Chapter of the Association for Computational Linguistics, EACL 2014 (2014)

Sentiment Analysis in Twitter for Spanish

Ferran Pla and Lluís-F. Hurtado

Universitat Politècnica de València
Camí de Vera s/n, 46022 València, Spain
{fpla,lhurtado}@dsic.upv.es

Abstract. This paper describes a SVM-approach for Sentiment Analysis (SA) in Twitter for Spanish. This task was part of the TASS2013 workshop, which is a framework for SA that is focused on the Spanish language. We describe the approach used, and we present an experimental comparison of the approaches presented by the different teams that took part in the competition. We also describe the improvements that were added to our system after our participation in the competition. With these improvements, we obtained an accuracy of 62.88% and 70.25% on the SA test set for *5-level* and *3-level* tasks respectively. To our knowledge, these results are the best results published until now for the SA tasks of the TASS2013 workshop.

Keywords: Sentiment Analysis, Twitter, Machine Learning.

1 Introduction

Twitter has become a popular micro-blogging site in which users express their opinions on a variety of topics in real time. The nature of texts used in Twitter (ungrammatical sentences with a lot of emoticons, abbreviations, specific terminology, slang, etc.) poses new challenges for researchers in Natural Language Processing (NLP) to provide effective solutions for Sentiment Analysis (SA) in micro-blogging texts. Therefore, the usual techniques of NLP must be adapted to these characteristics of the language, and new approaches must be proposed in order to successfully address this problem. NLP tools like POS taggers, parsers, or Named Entity Recognition (NER) tools usually fail when processing tweets because they generally are trained on grammatical texts and they perform poorly in micro-blogging texts.

Most of the work of SA on Twitter is for the English language and this is also true for the resources and tools available for NLP. Therefore, the TASS2013 workshop aims to be a framework for SA and on-line reputation analysis that is focused on the Spanish language. The organization of TASS2013 provided a corpus of Spanish tweets, *The General Corpus* [13], which is annotated with their polarity. This is a very important resource that allows researchers to compare their approaches for the SA problem on Twitter by using the same data.

SA has been widely studied in the last decade in multiple domains. Most work focuses on classifying the polarity of the texts as positive, negative, mixed, or

E. Métais, M. Roche, and M. Teisseire (Eds.): NLDB 2014, LNCS 8455, pp. 208–213, 2014.

neutral. The pioneering works in this field used supervised [7] or unsupervised (knowledge-based) [12] approaches. In [7], the performance of different classifiers on movie reviews was evaluated. In [12], some patterns containing POS information were used to identify subjective sentences in reviews to then estimate their semantic orientation. The construction of polarity lexicons is another widely explored field of research. Opinion lexicons have been obtained for English [3] [15] and also for Spanish [8].

Research works about SA on Twitter are much more recent. Twitter appeared in the year 2006 and the early works in this field are from 2009 when Twitter started to achieve popularity. Some of the most significant works are [1], [2], and [5]. A survey of the most relevant approaches to SA on Twitter can be see in [4] and [14]. The SemEval2013 competition has also dedicated a specific task for SA on Twitter [16], which shows the great interest of the scientific community.

In this work we present our approach and the results obtained for the SA tasks proposed at the TASS2013 workshop. Two different sub-tasks called *5-level* and *3-level* were proposed. Both sub-tasks differ only in the polarity granularity considered. The *5-level* sub-task uses N and $N+$ labels for negative polarity, P and $P+$ labels for positive polarity, and the *NEU* label for neutral polarity. The *3-level* sub-task only has three polarity levels: N, P, and *NEU*. In both sub-tasks, an additional label (*NONE*) was used to represent tweets with no polarity at all (objective tweets). The data used at TASS2013 workshop contains approximately 68000 Twitter messages (*tweets*) written in Spanish (between November 2011 and March 2012) by about 150 well-known personalities of the world of politics, economy, communication, mass media, and culture. Each tweet includes its ID, the creation date, and the user ID. The corpus is encoded in XML and it is divided into two sets: training (about 10%, 7219 tweets) and test (about 90%, 60798 tweets). The distributions per polarity of the training set is: 18.49% for N, 11.73% for N+, 9.28% for NEU, 20.54% for NONE, 17.07% for P, and 22.88% for P+.

2   System Description

The SA system proposed consists of 3 modules. The first module is the Preprocessing module, which performs the tokenization, lemmatization, and NER of the input tweet. A lemma reduction and a POS tagging process is also carried out in this module. The second module is the Feature Extraction module, which selects the features from the pre-processed tweet and obtains a feature vector. Some features require the use of a polarity lexicon of lemmas and words. To determine the best features, a tuning process is required during the training phase. The third module is the Polarity Classifier module, which uses a classifier to assign a polarity label to the tweet.

Before addressing the SA tasks, it is necessary to make a proper tokenization of the tweets that make up the training corpus and test corpus. Although there are a lot of tokenizers available on the web, they need to be adapted in order to address the tokenization of a tweet. In our system, we decided to use and adapt available tools for tokenization, lemmatization, NER, and POS tagging. We adapted the package *Tweetmotif* that is described in [5] to process Spanish tweets. We also used *Freeling* [6] (with the appropriate modifications for handling Twitter messages) for stemming, NER, and POS tagging. We added some functions to group special tokens into sigle tokens (e.g., *hashtags, web* addresses, *url, dates, numbers*, and some *punctuation* marks).

The SA task was addressed as a classification problem that consisted of determining the polarity of each tweet. We used WEKA, which is a tool that includes (among other utilities) a collection of machine-learning algorithms that can be used for classification tasks. Specifically, we used a SVM-based approach because it is a well-founded formalism, that has been successfully used in many classification problems. In the SA task, SVM has shown it ability to handle large feature spaces and to determine the relevant features. We used the NU-SVM algorithm [11] from an external library called *LibSVM*, which is very efficient software for building SVM classifiers. It is easy to integrate this software with WEKA thus allowing us to use all of WEKA's features. We used the *bag_of_words* approach to represent each tweet as a feature vector that contains the occurrences of the selected features.

The tuning process carried out had two objectives: to choose which features to include in the model, and to perform the parameter estimation of the SVM. We conducted this optimization by means of a 10-fold cross validation using the official TASS2013 training set as the development set. In order to determine the features of the model, the following four parameters were considered: *lemma frequency (f)*, *bilemma*, *selPOS*, and *polarity lexicon (DIC)*. The *lemma frequency (f)* parameter determines the minimum frequency necessary to consider a lemma as a feature. The *bilemma* parameter determines if bigrams of lemmas (in addition to single lemmas) are included as features in the model. The *selPOS* parameter determines if only the lemmas that belong to a prefixed set of POS are included in the model. When *selPOS* is used, only those lemmas belonging to *nouns, verbs, adjectives, adverbs* POS (in addition to the *emoticons* and *exclamations*) are included in the model. Finally, *DIC* determines if external polarity lexicons are used. The external lexicons used by our system in the TASS2013 competition [9] were lists of words and lemmas with their a priori polarity. One of the lexicons used was originally for English [15] that was translated into Spanish automatically, and other [8] lexicon was a list of words that was originally in Spanish. With these two resources we constructed our original dictionary *(DIC)*. Then, we combined *DIC* with the lexicon presented in [10] in order to obtain an improved lexicon *(DIC-improved)*.

Table 1 shows the *Accuracy* and confidence interval (with a 95% level of confidence) of the 10-fold cross validation process for *5-level* and *3-level* subtasks and for different combinations of the features under consideration (from system s_1 to system s_{14}). We also include the average number of features for each system.

The *Accuracy* results obtained by the different systems considered were not statistically significant in many cases. However, there was a great difference in

Table 1. *Accuracy* results for the tuning process using the training set (10-fold)

System	Features	# features	Accuracy (%)	
			5-level	3-level
s_1	f=1	11436.7	45.41 ± 1.14	62.35 ± 1.33
s_2	f=1+selPOS	11308.7	44.99 ± 1.32	60.58 ± 1.17
s_3	f=1+DIC	11438.7	46.74 ± 1.06	65.16 ± 1.43
s_4	f=1+selPOS+DIC	11310.7	46.38 ± 1.05	64.37 ± 1.22
s_5	f=1+DIC-improved	11438.7	49.84 ± 1.23	68.17 ± 1.44
s_6	f=1+bilemma+DIC-improved	64686.7	49.63 ± 1.01	67.53 ± 1.08
s_7	f=1+selPOS+DIC-improved	11310.7	**50.20 ± 1.55**	67.20 ± 1.26
s_8	f=2	4533.0	45.93 ± 1.16	62.47 ± 1.07
s_9	f=2+DIC-improved	4535.0	50.12 ± 1.24	**68.35 ± 1.28**
s_{10}	f=2+bilemma+DIC-improved	16153.3	49.61 ± 1.37	68.04 ± 0.91
s_{11}	f=2+selPOS+DIC-improved	4410.0	50.09 ± 1.27	67.47 ± 1.30
s_{12}	f=3+DIC-improved	3015.3	49.85 ± 1.90	67.95 ± 1.41
s_{13}	f=3+bilemma+DIC-improved	9049.0	49.40 ± 1.29	67.78 ± 1.08
s_{14}	f=3+selPOS+DIC-improved	2904.3	49.73 ± 1.53	67.16 ± 1.43

performance between the systems that did not use lexicons and those that did use them; especially those using the *DIC-improved* lexicon. A detailed analysis of the hits obtained by the systems showed that there was a different of up to 5% for the correctly labeled tweets even on systems with the same precision.

Taking this into account, we decided to combine the systems in order to take advantage of their complementarity. Several different combination methods were tested and no relevant differences in accuracy were found. Finally, we decided to use a majority voting scheme. Each tweet was classified by each system and the polarity that was chosen by the majority of the systems was the polarity definitively assigned to the tweet. If a tie occurred, the most frequent among the tied polarities (in the training set) was selected. In the experimental work conducted, all the possible combinations of systems were tested.

The best results was obtained by combining 2 systems. When we used the system *voting1* (by combining s_7 and s_{13} systems) we improved the accuracy from 50.20% to 50.45% for the *5-level* task. With the system *voting2* (by combining s_{11} and s_{13} systems) we improved from 68.35% to 68.68% for *3-level* task.

3   The Evaluation on the Test Set

A total of 13 teams participated in the TASS2013 SA task. Fifty-six runs were submitted for evaluation in the competition. The official results ranged from 61.6% to 13.5% (for *5-level* task) and from 66.3% to 38.8% (for *3-level* task). The best results were obtained by machine learning-based approaches. A detailed description of the different approaches is available on the TASS2013 website.

We constructed new models for the *5-level* and *3-level* tasks with the best set of features obtained in the tuning phase. We tested these models on the

Table 2. *Accuracy* results on the test set

System	Acurracy (%)	
	5-level	3-level
s_7	**60.02 ± 0.39**	68.85 ± 0.37
s_9	59.21 ± 0.39	**69.64 ± 0.37**
voting1	**62.88 ± 0.38**	70.16 ± 0.36
voting2	62.77 ± 0.38	**70.25 ± 0.36**
UA	61.62 ± 0.39	66.28 ± 0.38
ELHUYAR	60.10 ± 0.39	68.65 ± 0.37
UPV-ELiRF	57.60 ± 0.39	67.40 ± 0.37

test set supplied at the TASS2003 competition. The results obtained with the confidence interval are show in Table 2. It also include the 3 best approaches at the TASSS2013 competition: UA, ELHUYAR and UPV-ELiRF(our system).

Note that *Accuracy* results are higher than those obtained on the training set. This was true for all of the approaches presented at this competition. We have no clear explanation for this, it may be because the distribution of tweets by category in the training set (i.e, P+,22%; NONE,20%) is different from the test set (i.e, P+,34%; NONE,35%), or it may be because the process of manual supervision was different for these training and test sets. Our best voting systems outperform the ELHUYAR and UA systems for both the *3-level* and the *5-level* tasks with statistical significance. For the *3-level* task, our individual system s_9 also outperformed the other approaches with statistical significance.

4   Conclusions

In this paper, we have presented our approach for the SA task of the TASS2013 competition. For the classification stage, we used a Support Vector Machine approach with WEKA and the external LibSVM library.

We have presented the improvements we have made to the system that we submitted to the TASS2013 competition. These improvements consisted of adding new features to the classifiers, the construction of new polarity dictionaries, and the combination of different models by means of voting techniques. With these improvements, we obtained the best results for *5-level* and *3-level* tasks with an accuracy of 62.88% and 70.25% respectively. We think that the corpus and the gold standards provided at the TASS2013 competition (which are available on the TASS2013 webpage) and the evaluation presented in this work will be helpful for other research groups that are interested in the SA task.

As future work, we plan to continue working on this task, taking into account new features and resources. Specifically, these can include using more accurate text normalization techniques for improving POS tagging and NER for tweet domain and using and adapting new language resources for conducting a deep syntactic analysis to tackle some specific issues that could improve SA tasks in Twitter, such as negation, modifiers of polarity (adverbs), or coreference.

Acknowledgments. This work has been funded by the projects, DIANA (MEC TIN2012-38603-C02-01) and Tímpano (MEC TIN2011-28169-C05-01).

References

1. Barbosa, L., Feng, J.: Robust sentiment detection on Twitter from biased and noisy data. In: Proceedings of the 23rd International Conference on Computational Linguistics: Posters, Association for Computational Linguistics, pp. 36–44 (2010)
2. Jansen, B.J., Zhang, M., Sobel, K., Chowdury, A.: Twitter power: Tweets as electronic word of mouth. Journal of the American Society for Information Science and Technology 60(11), 2169–2188 (2009)
3. Liu, B., Hu, M., Cheng, J.: Opinion observer: Analyzing and comparing opinions on the web. In: Proceedings of the 14th International Conference on World Wide Web, WWW 2005, pp. 342–351. ACM, New York (2005)
4. Martínez-Cámara, E., Martín-Valdivia, M.T., Ureña-López, L.A., Montejo-Raéz, A.: Sentiment analysis in twitter. Natural Language Engineering 1(1), 1–28 (2012)
5. O'Connor, B., Krieger, M., Ahn, D.: Tweetmotif: Exploratory search and topic summarization for twitter. In: Cohen, W.W., Gosling, S. (eds.) Proceedings of the Fourth International Conference on Weblogs and Social Media, ICWSM 2010, Washington, DC, USA, May 23-26, The AAAI Press (2010)
6. Padró, L., Stanilovsky, E.: Freeling 3.0: Towards wider multilinguality. In: Proceedings of the Language Resources and Evaluation Conference (LREC 2012). ELRA, Istanbul (2012)
7. Pang, B., Lee, L., Vaithyanathan, S.: Thumbs up? sentiment classification using machine learning techniques. In: Proceedings of EMNLP, pp. 79–86 (2002)
8. Perez-Rosas, V., Banea, C., Mihalcea, R.: Learning sentiment lexicons in spanish. In: Chair, N.C.C., Choukri, K., Declerck, T., Doğan, M.U., Maegaard, B., Mariani, J., Odijk, J., Piperidis, S. (eds.) Proceedings of the Eight International Conference on Language Resources and Evaluation (LREC 2012). European Language Resources Association (ELRA), Istanbul (2012)
9. Pla, F., Hurtado, L.F.: Análisis de sentimientos en twitter. In: Proceedings of the TASS workshop at SEPLN 2013, IV Congreso Español de Informática (2013)
10. Saralegi, X., San Vicente, I.: Elhuyar at tass 2013. In: Proceedings of the TASS workshop at SEPLN 2013, IV Congreso Español de Informática (2013)
11. Schölkopf, B., Smola, A.J., Williamson, R.C., Bartlett, P.L.: New support vector algorithms. Neural Comput. 12(5), 1207–1245 (2000)
12. Turney, P.D.: Thumbs up or thumbs down? semantic orientation applied to unsupervised classification of reviews. In: ACL, pp. 417–424 (2002)
13. Villena-Román, J., García-Morera, J.: Workshop on sentiment analysis at sepln 2013: An over view. In: Proceedings of the TASS Workshop at SEPLN 2013, IV Congreso Español de Informática (2013)
14. Vinodhini, G., Chandrasekaran, R.: Sentiment analysis and opinion mining: A survey. International Journal 2(6) (2012)
15. Wilson, T., Hoffmann, P., Somasundaran, S., Kessler, J., Wiebe, J., Choi, Y., Cardie, C., Riloff, E., Patwardhan, S.: Opinionfinder: A system for subjectivity analysis. In: Proceedings of HLT/EMNLP on Interactive Demonstrations, Association for Computational Linguistics, pp. 34–35 (2005)
16. Wilson, T., Kozareva, Z., Nakov, P., Rosenthal, S., Stoyanov, V., Ritter, A.: Semeval-2013 task 2: Sentiment analysis in twitter. In: Proceedings of the International Workshop on Semantic Evaluation, SemEval, vol. 13 (2013)

Cross-Domain Sentiment Analysis Using Spanish Opinionated Words*

M. Dolores Molina-González, Eugenio Martínez-Cámara,
M. Teresa Martín-Valdivia, and L. Alfonso Ureña-López

Computer Science Department,
University of Jaén,
Campus Las Lagunillas, 23071, Jaén, Spain
{mdmolina,emcamara,maite,laurena}@ujaen.es

Abstract. A common issue of most of NLP tasks is the lack of linguistic resources in languages different from English. In this paper is described a new corpus for Sentiment Analysis composed by hotel reviews written in Spanish. We use the corpus to carry out a set of experiments for unsupervised polarity detection using different lexicons. But, in addition, we want to check the adaptability to a domain for the lists of opinionated words. The obtained results are very promising and encourage us to continue investigating in this line.

Keywords: Sentiment Polarity Detection, Spanish Opinion Mining, Spanish hotel review corpus, domain adaptation.

1 Introduction

Sentiment Analysis (SA), also known as Opinion Mining (OM) is a challenging task that combines Natural Language Processing (NLP) and Text Mining (TM). Polarity classification is one of the most studied tasks of OM that is focused on determining which is the overall sentiment-orientation of the opinions contained within a given document. The document is supposed to contain subjective information such as product reviews or opinionated posts in blogs. In this paper, we focus on semantic orientation for polarity classification in reviews over a tourism domain. We want to analysis the goodness of some lexicons and the adaptability to a specific domain. Specifically we have chosen the tourism domain to carry out our experimental study. We have generated a corpus with hotel reviews from the Tripadvisor website[1]. The results obtained are very promising being even comparable with machine learning approach.

On the other hand, although SA is a very recent area, the number of papers, applications and resources dedicated to this task is impressive. However, most

* This work has been partially supported by a grant from the Fondo Europeo de Desarrollo Regional (FEDER), ATTOS project (TIN2012-38536-C03-0) from the Spanish Government. The project AORESCU (TIC - 07684) from the regional government of Junta de Andaluca partially supports this manuscript.

[1] http://www.tripadvisor.es

E. Métais, M. Roche, and M. Teisseire (Eds.): NLDB 2014, LNCS 8455, pp. 214–219, 2014.
© Springer International Publishing Switzerland 2014

of works related to opinion mining only deal with English texts. In this paper we focus on Spanish SA. Our main interest is to check the behaviour of different Spanish lexicons generated using as a base the SOL (Spanish Opinion Lexicon) resource [1]. This lexicon was manually checked obtaining the iSOL (improved SOL) resource.

The rest of the paper is organised as follows: The next section presents related studies that apply a semantic orientation approach focusing mainly on the use of lexical resources. Section 3 introduces the main resources used and describes the construction process and the features of a corpus of hotel reviews (COAH) that we used in our experiments. Section 4 presents the results obtained in the experiments we performed. Finally, the conclusions and future work are presented.

2   Related Works

Several studies have been published related to extract the opinion from the users reviews posted in touristic web sites. In [2] a supervised polarity classification system is described. The authors compiled a corpus from the travel column of Yahoo!² , which is composed by 1191 reviews. The authors study the performance of two machine learning classifiers, Naïve Bayes and SVM, and the N-gram based character language model for SA. In this case SVM was the algorithm that reached the best results. Chinese hotel reviews have also been utilised by the SA research community. In [3] the authors classify Chinese hotel reviews. As in the former work, the authors develop a supervised classifier based on the use of the algorithm SVM.

A considerable number of papers have also been published proposing methodologies to adapt resources to the hotel domain. In [4] is proposed a method to build a domain-dependent polarity classification system. The domain selected by the authors is hotel reviews. Each review is represented by a set of domain-independent features and a set of domain-dependent ones. The domain-independent features are extracted from SentiWordNet. To build the set of domain-dependent features the authors propose to take the lexicon built by Hu and Liu [5] and choose those positive/negative words that occur in a significant number of positive/negative reviews of the training corpus used for the experimentation. In [6] is proposed a method to adapt a domain-independent sentiment linguistic resource, like SentiWordNet, to a specific domain. The assessment is done with a corpus of English hotel reviews downloaded from Tripadvisor [7].

Our proposal for the domain adaptation problem is the use of specific lists of opinion words per each domain. So, in the following sections a corpus of Spanish hotel reviews and an opinion list for tourism domain are described. Also, a set of experiments are shown with the aim of illustrating the value of these new linguistic resources.

---

² http://travel.yahoo.com/

3   Resources: Corpora and Word Lists

The development of new linguistic resources is very important to make progress in solving the problem of cross-domain SA. Also, the need of new linguistic resources is higher in languages other than English, like Spanish. So, the main contribution of this paper is the description of a new corpus of reviews in the tourism domain, and also new experiments that certificate the goodness of two sentiment lexicons developed by us.

Firstly, a Spanish corpus of hotel reviews has been compiled. The corpus is called COAH, which means Corpus of Opinion about Andalusian Hotels. Using COAH, an unsupervised polarity classification system has been developed with the aim of assessing two domain-independent opinion lexicons and a sentiment lexicon adapted to the tourism domain. The domain-independent lexicons are SOL and iSOL lexicons [1], and the tourism lexicon is eSOLHotel.

3.1   Corpus: COAH

For our experiments we have created the Corpus of Opinion about Andalusian Hotels COAH from the TripAdvisor site. The collection contains 1,835 reviews not written by professional writers, but rather by the web users. This may appear anecdotal, though it increases the difficulty of the task, because the texts may not be grammatically correct, or they can include spelling mistakes or informal expressions. We have selected only Andalusian hotels, ten hotels per each province of Andalusian (Almería, Cádiz, Córdoba, Granada, Jaén, Huelva, Málaga and Sevilla). We have selected ten hotels, five of them with higher rating and the other five with worse rating. All the hotels must have at least twenty opinions in the latter years written in Spanish. Finally, we have obtained 1,835 reviews.

The opinions are rated on a scale from 1 to 5. A rank of 1 means that the hotel is very bad, and 5 means very good. Rated 3 hotels can be categorised as "neutral" which means the user consider the hotel is neither bad nor good. Table 1 shows the number of reviews per rating. In our experiments we discarded the neutral reviews. In this way, opinions rated with 3 were not considered, the opinions with ratings of 1 or 2 were considered as negative reviews (519 in total) and those with ratings of 4 or 5 were considered as positives (1,025 in total).

Table 1. Rating distribution

Rating	1	2	3	4	5	Total
#Reviews	316	203	291	493	532	1835

In Table 2 is shown some interesting features of the corpus.

Table 2. COAH statistics

#Reviews	1,835
#Hotels	80
Mean of reviews per hotel	22.93
#Tokens	266,410
#Sentences	10,042
#Adjectives	17,910
#Adverbs	15,357
#Verbs	38,889
#Names	53,924
Mean of tokens per sentence	26.52
Mean of tokens per review	145.18
Mean of adjectives per review	9.76
Mean of adverbs per review	8.36
Mean of verbs per review	21.19
Mean of names per review	29.38

3.2 iSOL and eSOLHotel Lexicons

The iSOL resource was generated from the BLEL lexicon [5] by automatically translating it into Spanish and obtaining the SOL (Spanish Opinion Lexicon) resource. Then this resource was manually reviewed in order to improve the final list of words obtaining iSOL (improved SOL). The iSOL is composed of 2,509 positive and 5,626 negative words, thus in total the Spanish lexicon has 8,135 opinion words.

On the other hand, the eSOLHotel List is a resource generated from the iSOL lexicon by using domain knowledge. In this respect, we chose the Spanish part of the comparable SFU Reviews Corpus [8]. The SFU Reviews Corpus is composed of reviews of products in English and Spanish. The Spanish reviews are divided into eight categories: books, cars, computers, washing machine, hotels, movies, music and phones. In order to generate the enriched Spanish Lexicon for hotels (eSOLHotel), we search the most frequent words into the hotel category in the SFU Spanish Review Corpus.

4   Experiments and Results

In order to evaluate the experiments, we used the traditional measures employed in text classification: precision (P), recall (R), F1 and Accuracy (Acc.). The polarity of a review is calculated by taking into account the total number of positive words (#positive) and the total number of negative words (#negative) within the review. Table 3 shows the results obtained by using the three lists of opinionated words over the COAH corpus.

Table 3. Results obtained by using the three lists of opinionated words

	Macro-P	Macro-R	Macro-F1	Acc.
SOL	84.70%	75.22%	79.68%	82.24%
iSOL	91.61%	83.25%	87.23%	88.46%
eSOLHotel	91.59%	84.31%	87.80%	89.05%

4.1   Comparison with Other Related Work

In the literature we can find an interesting resource called the Spanish Emotion Lexicon (SEL) provided by Sidorov [9]. This resource is freely available for research purposes[3]. SEL is composed of 2,036 words that are associated with the measure of Probability Factor of Affective use (PFA) with respect to at least one basic emotion or category: *joy, anger, fear, sadness, surprise,* and *disgust.* In order to establish a feasible comparison by using the SEL resource for binary classification of COAH, we considered the joy and surprise categories as positive and the others as negative.

As is widely known, supervised learning overcomes unsupervised learning, so an unsupervised system is better when its results will be closer to the ones reached by a supervised system. Thus, a supervised experiment has been carried out with the aim of comparing the lexicon-based classifier described previously. The supervised system used is the same that is described in [10], i.e. the SVM algorithm has been used as classifier, and the reviews have been represented as a set of vectors of tokens weighted by TF-IDF. The results achived with ML and SEL are shown in Table 4.

Table 4. Comparison for binary classification of COAH by using ML (SVM algorithm) and eSOLHotel

	Macro-P	Macro-R	Macro-F1	Acc.
SEL	81.72%	69.00%	74.82%	78.16%
SVM	95.22%	93.14%	94.17%	94.82%
eSOLHotel	91.59%	84.31%	87.80%	89.05%

The high results in Table 3 show the validity of the two lexicons presented in this paper, iSOL and eSOLHotel, for polarity classification of Spanish reviews in hotel domain. Besides, in the same table we can observe that accuracy is better with eSOLHotel, lexicon where we are implement the domain knowledge.

Therefore, we consider that the lexicons developed and the new corpus COAH, which are freely available, are valuable resources for the Spanish SA research community.

Regarding Accuracy and F1 of Table 3 and Table 4, the percentage difference between eSOLHotel and SVM is only 6.27% and 7% respectively. The reduced

---

[3] http://www.cic.ipn.mx/~sidorov/#SEL

difference shows the goodness of eSOLHotel, which is also certificated by the good results reached with the hotel reviews section of the corpus SFU. This good performance also shows that eSOLHotel covers correctly vocabulary related to tourism specially the vocabulary related to hotels.

5 Further Work

Currently we are working on the development of several Spanish lexicons for domain-dependent SA following the method proposed here, i.e. selecting the words with a higher frequency in a corpus. Also, we want to deep in the evaluation of the domain-dependent lists of opinion words using larger sets of reviews.

References

1. Molina-González, M.D., Martínez-Cámara, E., Martín-Valdivia, M.T., Perea-Ortega, J.M.: Semantic orientation for polarity classification in spanish reviews. Expert Systems with Applications 40(18), 7250–7257 (2013)
2. Ye, Q., Zhang, Z., Law, R.: Sentiment classification of online reviews to travel destinations by supervised machine learning approaches. Expert Syst. Appl. 36(3), 6527–6535 (2009)
3. Shi, H.X., Li, X.J.: A sentiment analysis model for hotel reviews based on supervised learning. In: 2011 International Conference on Machine Learning and Cybernetics (ICMLC), vol. 3, pp. 950–954 (July 2011)
4. Dehkharghani, R., Yanikoglu, B., Tapucu, D., Saygin, Y.: Adaptation and use of subjectivity lexicons for domain dependent sentiment classification. In: 2012 IEEE 12th International Conference on Data Mining Workshops (ICDMW), pp. 669–673 (December 2012)
5. Hu, M., Liu, B.: Mining and summarizing customer reviews. In: Proceedings of the Tenth ACM SIGKDD International Conference on Knowledge Discovery and Data Mining, KDD 2004, pp. 168–177. ACM, New York (2004)
6. Demiroz, G., Yanikoglu, B., Tapucu, D., Saygin, Y.: Learning domain-specific polarity lexicons. In: 2012 IEEE 12th International Conference on Data Mining Workshops (ICDMW), pp. 674–679 (December 2012)
7. Wang, H., Lu, Y., Zhai, C.: Latent aspect rating analysis on review text data: A rating regression approach. In: Proceedings of the 16th ACM SIGKDD International Conference on Knowledge Discovery and Data Mining, KDD 2010, pp. 783–792. ACM, New York (2010)
8. Taboada, M., Grieve, J.: Analyzing appraisal automatically. Technical report, Stanford University (March 2004)
9. Sidorov, G., Miranda-Jiménez, S., Viveros-Jiménez, F., Gelbukh, A., Castro-Sánchez, N., Velásquez, F., Díaz-Rangel, I., Suárez-Guerra, S., Treviño, A., Gordon, J.: Empirical study of machine learning based approach for opinion mining in tweets. In: Batyrshin, I., González Mendoza, M. (eds.) MICAI 2012, Part I. LNCS, vol. 7629, pp. 1–14. Springer, Heidelberg (2013)
10. Martínez-Cámara, E., Martín-Valdivia, M.T., Perea-Ortega, J.M., Ureña-López, L.A.: Opinion classification techniques applied to a spanish corpus. Procesamiento del Lenguaje Natural 47, 163–170 (2011)

Real-Time Summarization of Scheduled Soccer Games from Twitter Stream

Ahmed A.A. Esmin, Rômulo S.C. Júnior, Wagner S. Santos,
Cássio O. Botaro, and Thiago P. Nobre

Department of Computer Science, Federal University of Lavras, MG, Brazil
ahmed@dcc.ufla.br, romulojr@sistemas.ufla.br,
{wagnersouza,cassiobotaro,thiago_pnobre}@comp.ufla.br

Abstract. This paper presents the real-time summarization of scheduled soccer games from flows of Twitter stream. During events, many messages (tweets) are sent describing and expressing opinions about the game. The proposed approach shrinks the stream of tweets in real-time, and consists of two main steps: (i) the sub-event detection step, which determines if something new has occurred, and (ii) the tweet selection step, which picks a few representative tweets to describe each sub-event. We compare the automatic summaries generated in some of the soccer games of Brazilian, Spanish and England (2013-2014) Leagues with the live reports offered by ESPN! Sports, Globo Esporte, Yahoo! Sports and livematch.Com web site. The results show that the proposed approach is efficient and can produce real-time summarization with good quality.

Keywords: Real-time Summarization, Twitter Stream, Social Media.

1 Introduction

Twitter is a microblog service where users post messages ("tweets") of no more than 140 characters. With over 500 million active users[1] [1].

The community of users live tweeting about a given event generates rich contents describing sub-events that occur during an event (e.g., goals or penalties in a soccer game). All those users share valuable information providing live coverage of events [2,3]. However, this huge amount of information makes difficult for the user: (i) to follow the full stream on specific event while finding out about new sub-events, and (ii) to retrieve from Twitter the main, summarized information about which are the key things happening at the event. In the context of exploring the potential of Twitter as a means to follow an event, we address the task of summarizing Twitter contents by providing the user with a summed up stream that describes the key sub-events of the soccer game.

---

[1] http://www.mediabistro.com/alltwitter/
500-million-registered-users_b18842

E. Métais, M. Roche, and M. Teisseire (Eds.): NLDB 2014, LNCS 8455, pp. 220–223, 2014.
© Springer International Publishing Switzerland 2014

In this paper, we discuss an effort to automatically create "summary" posts of Twitter trending scheduled events and perform experiments on scheduled soccer games, where the start time is known [3].

The rest of the paper is organized as follows: Section 2 presents the related works, Section 3 presents the dataset, Section 4 describes the proposed approach for real-time summarization, Section 5 presents the experimental results. Finally, the conclusions and future works are given in Section 6.

2   Related Work

Automatic summarization of events from tweets is a new area of research. Prior work on micro-blog summarization includes summarization of a set of status updates by Sharifi et al. [4,5] and event summarization by Chakrabarti et al. [6]. Some have tackled the task in an off-line mode, after the events were finished. For instance, Hannon et al. [7] present an approach for the automatic generation of video highlights for soccer games after they finished. Others, such as Petrovic et al. [8], have shown the potential of Twitter for the detection and discovery of events from tweets. Chakrabarti et al. [6] use Twitter to generate summaries of long running, structure rich events, where multiple events share the same underlying structure.

3   The Dataset

For the design and evaluation of our approach, we collected a dataset of tweets sent during the games of a soccer competition from Twitter regarding the Brazilian 2013 First Division Soccer League, the First Division Spanish league - La Liga (2013-2014) and the English Premier League (2013-2014). We also collect twitters during the 2013 FIFA Confederations Cup, which was held in Brazil. Tweets for this dataset were recorded through Twitter's Streaming API using a track stream that receives tweets based on specific keyword query and # tags [9]. Table 1, shows some of the games were selected randomly from the set of games for the case study.

4   Real-Time Soccer Games Summarization

Sporting events consist of a sequence of moments, each of which may contain different actions by players, etc. Our summarization system relies on two properties of the Twitter stream as described by [5] and can be seen in Figure 1: (1) Sudden increases, or "spikes," in the volume of tweets in the stream suggest that something important just happened. (2) A high percentage of the contents of the tweets at a "spike" in volume contain text describing what happened.

To generate the summarization, we define a two-step process that enables to report information about new sub-events in different languages and then generate a set of few tweets as a summary of the game (the summarization result). The first step is to identify at all times, whether or not a new sub-event occurred in the last few seconds (or in predefined interval of time). The result of the first step determining if something new (sub-event) occurred; if so, the second step will take place and chosen the representative few of tweets that describes the sub-event in the language chosen by the user. During the game and at the end, this process will provide a set of tweets as a summary of the soccer game.

Table 1. List of games and the number of the collected Tweets

Num.	The Game	League	Number of Tweets
1	Atletico-MG vs. Atletico-PR	Brazilian league	2377
5	Brasil vs. espanha	Conf. Cup	7948
7	Real Madrid x Bilbao	La Liga	8319

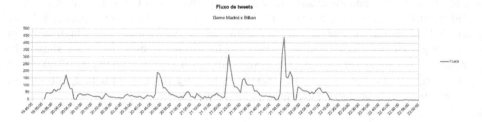

Fig. 1. Sample histogram of tweeting rates for a soccer game (Brazil vs Spain), where several peaks can be seen

Table 2. The results of the summarization game (Athletic Bilbao vs Real Madrid – 03 Feb 14)

Sub-events	Tweet Summarization (our system)	Narrator From the Livmatch.com
Game start	Now time for Athletic Bilbao v Real Madrid #HalaMadrid #LaLiga	0 We are under way in Bilbao!
Yellow card	Xabi Alonso Yellow Card : Athletic Bilbao 0-0 Real Madrid	¡HASTA EL FINAL, VAMOS REAL! #HALAMADRID	21 First yellow card of the match goes to Xabi Alonso after the Madrid man goes in with the studs up slightly on De Marcos. Probably deserved!
……..	………….	……………..

5   Evaluation of the Summarization Results

The case study games (Table 2) summaries results were manually evaluated by comparing them to reference live reports offered by the livematch website. In the manual evaluation process, each tweet in a system summary is classified as correct if it can be associated to a sub-event in the reference and is descriptive enough. The system achieved good precision results in the main sub-events like the Goal with more than 90%, Red cards (85%) and the Penalties (87%).

6   Conclusion and Future Work

We have presented summarization approach that use two steps applied to soccer games. Our system can detect the occurrence of sub-events from the tweeter stream and select a representative few tweets. We apply our approach on games on different soccer leagues with different languages and international 2013 FIFA Confederations Cup. The case study results show that the proposed approach is efficient and can produce real-time summarization with good quality. Our system generates real-time summaries with precision of 80.27% when compared to manually built reports from live reports offered journals.

As future work, we intend to evaluate the performance of the method on other types of sport events, Live TV shows, etc.

Acknowledgment. We would like to thank FAPEMIG, CAPES and CNPq (Brazilian agencies) for partial financial support.

References

1. Mishaud, E.: Twitter: Expressions of the whole self. Master's thesis, Department of Media and Communications, University of London (2007)
2. Becker, H., Iter, D., Naaman, M., Gravano, L.: Identifying content for planned events across social media sites. In: Proceedings of the Fifth ACM International Conference on Web Search and Data Mining (WSDM 2012), pp. 533–542 (2012)
3. Zubiaga, A., Spina, D., Amigó, E., Gonzalo, J.: Towards real-time summarization of scheduled events from twitter streams. In: Proceedings of the 23rd ACM Conference on Hypertext and Social Media (HT 2012), pp. 319–320. ACM, NY (2012)
4. Sharifi, B., Hunton, M.A.: andKalita, J. Summarizing Microblogs Automatically. In: Proc. ACL-HLT 2010 (2010)
5. Sharifi, B., Hunton, M.A., Kalita, J.: Experiments in Microblog Summarization. In: IEEE Second International Conference on Social Computing (2010)
6. Chakrabarti, D., Punera, K.: Event Summarization using Tweets. In: Proc. ICWSM 2011 (2011)
7. Hannon, J., McCarthy, K., Lynch, J., Smyth, B.: Personalized and automatic social summarization of events in video. In: Proceedings of the 16th International Conference on Intelligent User Interfaces (IUI 2011), pp. 335–338. ACM (2011)
8. Petrovi´c, S., Osborne, M., Lavrenko, V.: Streaming first story detection with application to twitter. In: Proc. of NAACL 2010 (2010)
9. Esmin, A.A.A., de Oliveira, R.L., Matwin, S.: Hierarchical Classification Approach to Emotion Recognition in Twitter, Machine Learning and Applications (ICMLA). In: 2012 11th International Conference on Machine Learning and Applications, December 12-15, vol. 2, pp. 381–385 (2012)

Focus Definition and Extraction
of Opinion Attitude Questions

Amine Bayoudhi[1], Hatem Ghorbel[2], and Lamia Hadrich Belguith[1]

[1] ANLP Group, MIRACL Laboratory, University of Sfax, B.P. 1088, 3018, Sfax Tunisia
bayoudhi.amine@gmail.com,
l.belguith@fsegs.rnu.tn
[2] ISIC Lab, HE-Arc Ingénierie, University of Applied Sciences, CH-2610 St-Imier Switzerland
hatem.ghorbel@he-arc.ch

Abstract. In Question Answering Systems (QAS), Question Analysis is an important task that consists in general in identifying the semantic type of the question and extracting the question focus. In this context, and as part of a framework aiming to implement an Arabic opinion QAS for political debates, this paper addresses the problem of defining the focus of opinion attitude questions and proposes an approach for extracting it. The proposed approach is based on semi-automatically constructed lexico-syntactic patterns. Evaluation results are considered very encouraging with an average precision of around 87.37%.

Keywords: Question Answering Systems, opinion question analysis, question focus extraction, opinion attitude questions, lexico-syntactic patterns.

1   Introduction

In Question Answering Systems (QAS), question analysis is an important task that attempts to determine the object of the question and the best approach to answering it [1]. It consists in identifying the semantic type of the question and subsequently inferring the type of the answer. However, this step is not always sufficient for finding answers as it might not say much about the query itself. Therefore, it is eventually followed by a focus extraction step [2].

In this paper, we investigate the improvement of the question analysis task by extracting the question focus, after having implemented the question type identification component in [3]. The current research is part of a framework aiming to implement an Arabic opinion QAS for political debates. More specifically, we present in this article a formal definition of the focus of the opinion attitude question, and we propose an original approach to extract it by means of lexico-syntactic patterns.

The rest of this paper is organized as follows. In section 2, we review a selection of previous works related to the focus definition and extraction task. In section 3, we detail our approach to define and to extract the opinion attitude question focus. In section 4, we describe and discuss the conducted experiments. Finally, we conclude and provide some perspectives in section 5.

E. Métais, M. Roche, and M. Teisseire (Eds.): NLDB 2014, LNCS 8455, pp. 224–227, 2014.
© Springer International Publishing Switzerland 2014

2   Related Work

The focus is an important informational component in a factoid question, used to help extracting answers [4]. Moldovan et al. for example define the question focus as *"the main information required by the question"* [5]. Whereas, Damljanovic et al. consider that it is *"a word or a sequence of words which define the question and disambiguate it by indicating what the question is looking for"* [2]. The main question that we want to evoke here is: are the definitions of factoid question focus available also to opinion question? If no, how can we define opinion question focus?

To the best of our knowledge, prior studies in opinion QAS did not bring much contribution to this issue. For example, Ku et al. have represented the opinion question focus by *"a set of content words"* and have extracted it by removing non-significant words from the question [6]. Other works such as those of Wiegand et al. [7] and Li et al. [8] did not even evoke the notion of question focus. A possible explanation for this might be that many of these opinion QAS were originally implemented to answer to factoid questions, and were adapted later to opinion questions. Proposed approaches for Focus extraction in opinion QAS rely mainly on Named Entity Recognition [9] and on eliminating non-significant words [6] such as interrogative words, stop words, opinion operators and negation words.

3   Proposed Approach

3.1   Definition of the Opinion Question Focus

According to Liu et al. [10], an opinion expression is composed of four basic opinion attributes (or components) represented as a quadruplet (g, so, h, t), where g is a target, so is the sentiment value of the opinion from opinion holder h on target g at time t (so is positive, negative or neutral, or a rating score), h is an opinion holder, and t is the time when the opinion is expressed. Since QAS work on separate questions not on continuous dialogues, the time component becomes in this context useless.

Similarly to opinion expressions, an opinion question is hence represented as a triplet (g, so, h). However, its particularity is that it holds missing opinion attributes which are expected to be provided in the answer. These missing attributes define indeed the question type; the question focus is hence defined accordingly. Thus, for the attitude question (**AT**), given an opinion question qt having the question type $opt="AT"$, the question focus will be represented by the triplet $(g, so:?, h)$ where g is a target, so is the sentiment value of the opinion from opinion holder h on target g, and h is an opinion holder. The question focus here is written according to the representation of Singer [11]. The notation ":?" is meant to convey that the questioner wants to know the value of the given opinion attribute. For example, the focus of the AT question "ما هو موقف راشد الغنوشي من الديمقراطية؟" (What is the attitude of Rached Al-Ghanouchi toward democracy?) can be represented by the holder (Rached Al-Ghanouchi), the target (democracy) and the unknown semantic orientation.

3.2   Extraction of Opinion Question Focus: Case of Attitude Question

Our approach to extract the focus of the opinion AT question consists in applying lexico-syntactic patterns. It is based on a semi-automatic method for pattern building and a manual validation process. This method uses as learning corpus 120 AT questions extracted from the COPARQ corpus [12], and as Arabic syntactic parser the Stanford Parser [13]. In addition, it uses a set of extraction rules manually established in order to locate each opinion component on the basis of the pattern structure. We illustrate in Figure1 an example of lexico-syntactic pattern and the corresponding extraction rules.

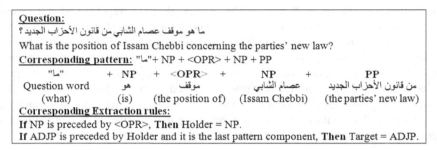

Fig. 1. Example of an AT question and its corresponding pattern

Regarding the manual validation process, it is performed by a linguistic expert who is responsible for detecting pattern ambiguity cases and resolving them. Pattern ambiguity could happen when we have more than one extraction rule associated to the same pattern. Resolving ambiguity is manually accomplished by adding more specified lexical information to the pattern.

After accomplishing the above steps, we have obtained at the end 23 lexico-syntactic patterns out of 120 questions. To each pattern, a unique extraction rule is associated. This rule will allow us to extract the holder and the target of the AT question.

4   Experiments and Discussions

Experiments were carried on by applying a 5-fold cross validation evaluation on the learning corpus dataset. In each time, 4 folds are allocated for the pattern extraction task and 1 fold for the test. Evaluation results were considered very encouraging with an average precision of around 87.37%.

Nevertheless, regarding that our approach relies strongly on the syntactic analysis, we believe that the obtained result was severely affected by the performance of the parser. In fact, although Stanford parser has achieved a good performance on the development test data sets[1], syntactic analysis results in our experimental corpus were less good. Apart from the parser performance, we can explain the obtained results by the fact that some AT questions are expressed without holder or target component.

---

[1] http://nlp.stanford.edu/software/parser-arabic-faq.shtml

5   Conclusions and Perspectives

This paper has investigated the definition of opinion attitude question focus and proposed an original approach for extracting the focus of opinion AT questions. This approach is based on semi-automatically constructed lexico-syntactic patterns. The purpose is to enhance the question analysis task in an Arabic opinion QAS for political debates, as well as to support the query construction step. Conducted experiments show that the focus extraction task has achieved over **87%** as an average precision.

As for our perspectives, we intend to extend the learning corpus and the proposed focus extraction method to include holder and target opinion questions. This is can be performed by enriching the pattern set to handle these two question types and building a polarity classification module to consider the semantic orientation of these questions.

References

1. Lally, A., Prager, J.M., McCord, M.C., Boguraev, B.K., Patwardhan, S., Fan, J., Fodor, P., Chu-Carroll, J.: Question analysis: How Watson reads a clue. IBM J. Res. & Dev. 56(3/4) (May/July 2012)
2. Damljanovic, D., Agatonovic, M., Cunningham, H.: Identification of the Question Focus: Combining Syntactic Analysis and Ontology-based Lookup through the User Interaction. In: LREC 2010 (2010)
3. Bayoudhi, A.: Classification des questions d'opinion dans un système de Questions-Réponses pour les débats politiques. In: CORIA 2013, Neuchâtel, Suisse, Avril 3-5, pp. 3–5 (2013)
4. El-Ayari, S.: Évaluation transparente de systèmes de questions-réponses: application au focus. In: RECITAL 2007, Toulouse, France, Juin 5-8 (2007)
5. Moldovan, D., Harabagiu, S.M., Pasca, M., Mihalcea, R., Girju, R., Goodrun, R., Rus, V.: The structure and performance of an open-domain question answering system. In: The 38th Meeting of the Association for Computational Linguistics (ACL 2000), pp. 563–570 (October 2000)
6. Ku, L.W., Liang, Y.T., Chen, H.H.: Question Analysis and Answer Passage Retrieval for Opinion Question Answering Systems. International Journal of Computational Linguistics and Chinese Language Processing 13(3), 307–326 (2008)
7. Wiegand, M., Momtazi, S., Kazalski, S., Xu, F., Chrupała, G., Klakow, D.: The Alyssa System at TAC QA 2008. In: TAC 2008, NIST (2009)
8. Li, F., Zheng, Z., Tang, Y., Bu, F., Ge, R., Zhang, X., Zhu, X., Huang, M.: THU QUANTA at TAC 2008 QA and RTE track. In: Text Analysis Conference (TAC 2008), Gaithersburg, Maryland USA (November 2008)
9. Balahur, A., Lloret, E., Ferrández, O., Montoyo, A., Palomar, M., Muñoz, R.: The DLSIUAES Team's Participation in the TAC 2008 Tracks. In: Proceedings of the Text Analysis Conference 2008 Workshop, Washington, USA (2008)
10. Liu, B.: Sentiment Analysis and Opinion Mining, tutorial given. In: EACL 2012, Avignon, France, April 23-27 (2012)
11. Singer, M.: Answering questions about discourse. Discourse Processes 13, 261–277 (1990)
12. Bayoudhi, A., Ghorbel, H., Hadrich Belguith, L.: Question Answering System for dialogues: a new opinion question taxonomy. In: Larsen, H.L., Martin-Bautista, M.J., Vila, M.A., Andreasen, T., Christiansen, H. (eds.) FQAS 2013. LNCS, vol. 8132, pp. 67–78. Springer, Heidelberg (2013)
13. Green, S., Manning, D.M.: Better Arabic Parsing: Baselines, Evaluations, and Analysis. In: COLING (2010)

Implicit Feature Extraction for Sentiment Analysis in Consumer Reviews

Kim Schouten and Flavius Frasincar

Erasmus University Rotterdam,
PO Box 1738, NL-3000 DR Rotterdam,
The Netherlands
{schouten,frasincar}@ese.eur.nl

Abstract. With the increasing popularity of aspect-level sentiment analysis, where sentiment is attributed to the actual aspects, or features, on which it is uttered, much attention is given to the problem of detecting these features. While most aspects appear as literal words, some are instead implied by the choice of words. With research in aspect detection advancing, we shift our focus to the less researched group of implicit features. By leveraging the co-occurrence between a set of known implicit features and notional words, we are able to predict the implicit feature based on the choice of words in a sentence. Using two different types of consumer reviews (product reviews and restaurant reviews), an F_1-measure of 38% and 64% is obtained on these data sets, respectively.

1 Introduction

Every day a vast amount of consumer reviews are written on the Web, where customers express their opinions about a product or service [1]. Not only do they describe their general sentiment or attitude towards the product or service, oftentimes specific aspects or features of that product or service are discussed in great detail [5]. This leaves researchers and companies alike with a valuable source of information about consumer sentiment.

However, in order to achieve the fine grained information that is needed for such analyses, the various aspects, or features, of a product or service must be recognized in the text first. Examples of such features include 'price', 'service', parts of a product like 'battery', or different meals and ingredients for restaurants. In most cases, these features are literally mentioned in the text. However, this is not always the case, as demonstrated in the example below, which is taken from the product review data set [4]:

"I like my phones to be small so I can fit it in my pockets."

Evidently, the feature referred to here is the size of the product, even though the word 'size' is never mentioned. However, words like 'small' and 'fit' give away that the feature implied here is the product's size. Unfortunately, detecting the implicit features is not always this straightforward.

E. Métais, M. Roche, and M. Teisseire (Eds.): NLDB 2014, LNCS 8455, pp. 228–231, 2014.

"I love the fact I can carry it in my shirt or pants pocket and forget about it."

The above example is an actual sentence in the product review data set, and according to the available annotations, the implicit feature in this case is also its size, however, it is easy to see that weight would also have been a good candidate.

2   Related Work

Earlier works that focus on implicit feature extraction are [6], where implicit features are found using semantic association analysis based on Point-wise Mutual Information, and [3], which uses co-occurrence Association Rule Mining to link opinion words as antecedents to implicit features as consequents.

Instead of linking opinion words to implicit features, [7] constructs a co-occurrence matrix between notional words and explicit features, using these co-occurrences to imply a feature in a sentence when no explicit feature is present. For each feature f_i, a score is computed that essentially is the sum of the co-occurrence frequencies $c_{i,j}$ between that feature i and the words j in the sentence.

$$score_{f_i} = \frac{1}{v} \sum_{j=1}^{v} \frac{c_{i,j}}{o_j}, \tag{1}$$

where v is the number of words in the sentence, f_i is the ith feature for which $score$ is computed, w represents the jth word in the sentence, $c_{i,j}$ is the co-occurrence frequency of feature i and lemma j, and o_j is the frequency of lemma o in the data set.

3   Method

To deal with the two violated assumptions mentioned in Sect. 2, a small but significant change is made in the construction of the co-occurrence matrix: instead of explicit features, manually annotated implicit features are used. This results in direct co-occurrence data between words and the implicit features that are to be determined. This change renders the two violated assumptions irrelevant, but it also introduces a dependence on annotated data.

Furthermore, all words are considered as context for implicit features. However, some word categories might be more useful as context words than others. Therefore we investigate a Part-of-Speech filter in which all combinations of word categories (i.e., noun, verb, adjective, and adverb) are tested as context words.

4   Results Analysis

All evaluations are performed using 10-fold cross-validation, with all sentences without an implicit feature being removed from the test set. The evaluation metric is the F_1-measure. Since precision and recall do not differ that much, only F_1-measure is reported. The two data sets that are used are a set of product reviews [4] and a set of restaurant reviews [2]. Note that for the latter, the "miscellaneous" category is removed as it is not really an implicit feature. Testing all mentioned Part-of-Speech filters pointed to the combination of nouns (NN), verb (VB), and adjectives (JJ) as the most effective context to find implicit features. This was the case for both data sets. The performance metrics using this context are given in Table 1.

Table 1. Evaluation results on both data sets, denoted as F_1-measure

	original method		revised method	
	all words	only NN, VB, and JJ	all words	only NN, VB, and JJ
product data	0.14	0.14	0.32	0.38
restaurant data	0.21	0.21	0.57	0.64

In general, one can conclude that directly creating the co-occurrence matrix with implicit features instead of indirectly with explicit features is a good strategy. The performance gain is significant, which will offset the disadvantage of needing labeled data. In terms of overall performance, the revised algorithm works best with a Part-of-Speech filter that only allows nouns, verbs, and adjectives. Concerning data sets, the revised algorithm works best with the restaurant data, which is relatively large and has only five different implicit features to choose from. Using the product data results in the worst performance, due to its limited size and increased difficulty: it has more different implicit features than the restaurant data and less instances per unique implicit feature. This makes it hard to properly train the algorithm.

5   Conclusion

The detection of features from reviews is important when measuring consumer sentiment on a fine-grained level. Adding the detection of implicit features, while a difficult task because the features themselves do not appear in the sentence, can increase the overall coverage of an aspect-level sentiment analysis tool. Besides a base method [7], two main revisions were discussed and evaluated on two data sets [2,4].

There are two main conclusions that can be drawn from the performed evaluation. The first is that it is much better to count the co-occurrence frequency between annotated implicit feature and notional words than to count the co-occurrence frequency between explicit features and notional words. Since the number of implicit features is usually much smaller than the number of explicit features, this will greatly reduce the size of the co-occurrence matrix as well, yielding better performance in terms of system load and processing time. The only drawback would be that this method is more domain dependent, as annotations of implicit features are required to train the system (i.e., do the counting).

The second is that filtering which words are allowed as context from which the implicit feature is derived is indeed helpful, albeit only for the revised method. A combination of nouns, verbs, and adjectives turns out to be most informative to extract the right implicit feature.

Possible directions for future work might include an extension to deal with more than one implicit feature in a sentence. While this is arguably not useful for the product review data, roughly one sixth of the restaurant review sentences has more than one implicit feature, rendering this a good way of reducing the number of false negatives. Another option might be to introduce a weighting scheme for the co-occurrences where the co-occurrence with different words can be weighted differently, based on for example additional domain or world knowledge. This could, for example, be taken from structured data like ontologies.

Acknowledgments. The authors are partially supported by the Dutch national program COMMIT. We also would like to thank Sven van den Berg, Marnix Moerland, Marijn Waltman, and Onne van der Weijde, for their contributions to this research.

References

1. Feldman, R.: Techniques and Applications for Sentiment Analysis. Communications of the ACM 56(4), 82–89 (2013)
2. Ganu, G., Elhadad, N., Marian, A.: Beyond the Stars: Improving Rating Predictions using Review Content. In: Proceedings of the 12th International Workshop on the Web and Databases (WebDB 2009) (2009)
3. Hai, Z., Chang, K., Kim, J.: Implicit Feature Identification via Co-occurrence Association Rule Mining. In: Gelbukh, A.F. (ed.) CICLing 2011, Part I. LNCS, vol. 6608, pp. 393–404. Springer, Heidelberg (2011)
4. Hu, M., Liu, B.: Mining and Summarizing Customer Reviews. In: Proceedings of 10th ACM SIGKDD International Conference on Knowledge Discovery and Data Mining (KDD 2004), pp. 168–177. ACM (2004)
5. Liu, B.: Sentiment Analysis and Opinion Mining. Synthesis Lectures on Human Language Technologies, vol. 16. Morgan & Claypool (2012)
6. Su, Q., Xu, X., Guo, H., Guo, Z., Wu, X., Zhang, X., Swen, B., Su, Z.: Hidden Sentiment Association in Chinese Web Opinion Mining. In: Proceedings of the 17th International Conference on World Wide Web (WWW 2008), pp. 959–968. ACM (2008)
7. Zhang, Y., Zhu, W.: Extracting Implicit Features in Online Customer Reviews for Opinion Mining. In: Proceedings of the 22nd International Conference on World Wide Web Companion (WWW 2013 Companion), pp. 103–104. International World Wide Web Conferences Steering Committee (2013)

Towards Creation of Linguistic Resources for Bilingual Sentiment Analysis of Twitter Data

Iqra Javed[1], Hammad Afzal[1], Awais Majeed[2], and Behram Khan[3]

[1] National University of Sciences and Technology, Islamabad, Pakistan
[2] Bahria University, Islamabad, Pakistan
[3] University of Manchester, UK
iqra.mscs17@students.mcs.edu.pk, hammad.afzal@mcs.edu.pk,
awais.majeed@gmail.com, khanb@cs.man.ac.uk

Abstract. This paper presents an approach towards bi-lingual sentiment analysis of tweets. Social networks being most advanced and popular communication medium can help in designing better government and business strategies. There are a number of studies reported that use data from social networks; however, most of them are based on English language. In this research, we have focused on sentiment analysis of bilingual dataset (English and Roman-Urdu) on topic of national interest (General Elections). Our experiments produced encouraging results with 76% of tweet's sentiment strength classified correctly. We have also created a bi-lingual lexicon that stores the sentiment strength of English and Roman Urdu terms. Our lexicon is available at: https://sites.google.com/a/mcs.edu.pk/codteem/biling_senti

Keyterms: Bi-lingual Sentiment Analysis, Language Resources, Social Network Analysis.

1   Introduction

The facilities provided by internet such as e-mail, online surfing and most recent and effective of them, the social networks like twitter with over 500 million active users and 50 million posts a day [1] play a major role in transforming life styles of communities. Social networks, particularly Twitter, hold the potential to serve as opinion mining and forecasting tool. A lot of research has been carried out in this domain but most of it has been done using English as language of expression. The popularity of social media has also extended to developing countries like Pakistan, with 30 million internet users[1] where Twitter is ranked among top ten most famous websites of the country. The national language Urdu is spoken by more than 66 million people in the world which is mostly used in the form of Roman-Urdu (written using English alphabets) on social networks.

---

[1] http://ansr.io/blog/pakistan-market-trends-2013-online-mobile-social/

E. Métais, M. Roche, and M. Teisseire (Eds.): NLDB 2014, LNCS 8455, pp. 232–236, 2014.

This research addresses the problem of bi-lingual classification and sentiment analysis of topic focused tweets. The bi-lingual classification includes the mechanism that is devised to separate the English tweets from the Roman-Urdu Tweets. A bilingual sentiment strength lexicon (BSL) is created using SentiStrength[2], WordNet[3] and a Bi-lingual Glossary (BG)[4]. The collected tweets are then analyzed to measure the popularity of political parties by classifying the positive and negative sentiments expressed in tweets.

2   Literature Review

Manually marked lexical lists and machine learning techniques have been mostly used in sentiment analysis to analyze mood and emotion in tweets. [2] provide qualitative as well as quantitative analysis to explore opinion convergence using twitter data of Singapore General Election 2011. They used Sentistrength for polarity classification of dataset. The research concluded that informative tweets were more effective than the affective tweets. However, a context-specific vocabulary is required for analysis of every event based on the contextual differences. They performed experiments using only English Tweets. In another research, an English-Romanian-Russian lexical resource is created that is capable of providing six basic emotion types (anger, disgust, joy, surprise, fear and sadness) for sentiment analysis [3].

3   Method

A bilingual lexicon is created that provides the sentiment strength score to English as well as Roman-Urdu words. We used this lexicon to perform sentiment analysis on dataset (89,000 political event based tweets) composed of English and Roman-Urdu tweets. The proposed methodology comprises of three major steps. In first step, topic and language classifications are performed. In second step, a Bi-lingual Sentiment Lexicon (BSL) is semi-automatically created using the existing (Enlgish) Senti-Strength, WordNet and BG (see Section 3.2). In third step, tweets are analyzed and degree of sentiments is computed for each tweet in dataset using BSL.

3.1   Classification of Tweet

89,000 tweets are collected using keywords related to Pakistani political parties and major cities. Tweets are then classified as political (82,224) and non-political (6,847) based on a filteration method using stop-words. Furthermore, a language classification method is devised and applied on dataset to classify English and Roman-Urdu tweets (see Table 1 for details).

---

[2] http://sentistrength.wlv.ac.uk/
[3] http://wordnet.princeton.edu/wordnet/
[4] http://www.scribd.com/doc/14203656/
  English-to-Urdu-and-Roman-Urdu-Dictionary

Table 1. Language Classification using bilingual classifier

City	Number of English Tweets				Number of Roman-Urdu Tweets				Total Tweets
	PTI	PML	PPP	MQM	PTI	PML	PPP	MQM	
Islamabad	5586	544	2026	1969	1710	292	911	946	13984
Lahore	6319	4509	3513	5122	1706	1443	1161	1573	10128
Karachi	7149	2236	7190	7067	2167	1693	2851	2320	32673
Peshawar	4415	459	1219	1091	1675	662	360	247	25346
Queta	19	5	21	3	18	8	12	7	93

Language classification is carried out using a basic lookup method in WordNet. Each tweet is classified as English on the basis of presence of English words in it by using the formula

$$Weight_{tweet} = \frac{W_{eng}}{W_{Total}} * 100 \tag{1}$$

where, W_{eng} and W_{total} are number of English tokens in tweet and total number of tokens in tweets respectively. A detailed discussion on classifications (Political/Non-Political and Language) is provided in [4].

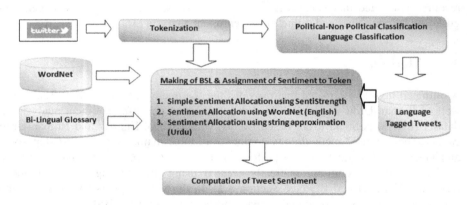

Fig. 1. Workflow of bilingual Sentiment Analysis

3.2   Construction of Bi-lingual Sentiment Lexicon (BSL) and Sentiment Allocation

A unified sentiment strength lexicon referred as Bilingual Sentiment Lexicon (BSL) is created which comprises of terms from English and Roman-Urdu with their sentiment strengths. Three resources are utilized in creation of BSL: SentiStrength, WordNet and a bilingual glossary. Sentiment strength in BSL ranges from -4 to 4, representing -4 as most negative and 4 as most positive strength. WordNet is utilized to enhance the coverage of Senti-strength by using the synonym lists based on word's conceptual-semantic relations. For Roman-Urdu tokens, relevant English token (English Translation) is retrieved from BG and is assigned sentiment score by using Senti-Strength or by using semantic association provided by WordNet (in case word is not in original senti-strength). A number of Roman-Urdu terms were still not found as typographical errors or variation in writing styles (transliteration variation). Bigram-Cosine similarity is applied in order to deal with such errors. Thus, BSL provides sentiment strength of 2600 English and 3200 Roman-Urdu words. Frequencies of terms are calculated and terms are assigned sentiment strength. The sentiment score of each tweet is then computed as follows:

$$\text{Tweet-Sentiment (TS)} = \frac{F_1 * S_1 + F_2 * S_2 + F_3 * S_3 + \ldots F_n * S_n}{n} \tag{2}$$

where F_1, F_2... F_n are the frequencies of the tokens appearing in a tweet; S_1, S_2 ... S_n are the sentiment strength of the corresponding token; n is the number of tokens in a given tweet. Using the sentiment score of individual tweets, the popularity score for each political party can then be computed as:

$$\text{Popularity-Score of Party (PSp)} = \sum_{i=0}^{n} \frac{TS_{pi}}{n} \tag{3}$$

where TS_{pi} is the strength of a tweet belonging to a particular political party p; n is the number of tweets belonging to party p.

4   Results and Evaluation

36,274 bilingual tokens are identified and assigned relative sentiment strength. For evaluation, three test sets are created from English, Roman-Urdu and randomly selected bi-lingual tweets. In all these test sets, tweets are analyzed manually to assign sentiment strength. The sentiment strengths of tweets in these selected test sets are then calculated our methodology and tweets are classified as either as positive or negative. Finally, results obtained are then analyzed for performance, using evaluation measures of recall, precision, accuracy and F-measure (as shown in Table.2).

Table 2. Performance Comparison of BSL on Multi-Lingual Test sets

Actual Class\Predicted class	Recall (%)	Precision (%)	F-Score (%)	Accuracy (%)
English Test Set	75	98	85	83
Roman-Urdu Test Set	63	96	76	69
Random Test Set	68	98	81	76

Our major focus (and research contribution) involves the sentiment allocation for Urdu tweets. Our system performed well in terms of precision (i.e. 96%); however, recall was lower due to low term identification rate because of transliteration variations. The recall can be improved by utilizing WordNet like lexicons for Urdu language (however, the work in this domain is still in infancy). Furthermore, the results obtained about political parties' popularities also slightly deviated from the actual results (mostly for cities with less internet using community) as limited society use social networks for opinion expression.

5   Conclusions

A lexicon based sentiment analysis approach is presented for analysis of short text messages such as tweets for English and Roman-Urdu. A bi-lingual lexicon is created that provides a list of English and Roman-Urdu terms along with their sentiment scores.

References

1. Measuring tweets, `http://blog.twitter.com/2010/02/measuring-tweets.html`
2. Wu, Y., Wong, J., Deng, Y., Chang, K.: An Exploration of Social Media in Public Opinion Convergence: Elaboration Likelihood and Semantic Networks on Political Events. In: Ninth IEEE International Conference on Dependable, Autonomic and Secure Computing (2011)
3. Bobicev, V., Maxim, V., Prodan, T., Burciu, N., Angheluş, V.: Emotions in Words: Developing a Multilingual WordNet-Affect. In: Gelbukh, A. (ed.) CICLing 2010. LNCS, vol. 6008, pp. 375–384. Springer, Heidelberg (2010)
4. Javed, I., Afzal, H.: Creation of bi-lingual Social Network Dataset using classifiers. Accepted in 10th International Conference on Machine Learning and Data Mining MLDM 2014, St. Petersburg, Russia, July 21-24 (2014)

Integrating Linguistic and World Knowledge for Domain-Adaptable Natural Language Interfaces

Hen-Hsen Huang[1], Chang-Sheng Yu[1], Huan-Yuan Chen[1], Hsin-Hsi Chen[1],
Po-Ching Lee[2], and Chun-Hsun Chen[2]

[1] Department of Computer Science and Information Engineering, National Taiwan University
No. 1, Sec. 4, Roosevelt Road, Taipei, 10617 Taiwan
[2] Telecommunication Laboratories, Chunghwa Telecom Co., Ltd.
No. 99, Dianyan Rd., Yangmei City, Taoyuan County, 32601 Taiwan, R.O.C.
{hhhuang,csyu}@nlg.csie.ntu.edu.tw,
{b00902057,hhchen}@ntu.edu.tw, {albertlee,jeffzpo}@cht.com.tw

Abstract. Nowadays, natural language interfaces (NLIs) show strong demands on various smart devices from wearable devices, cell phones, televisions, to vehicles. Domain adaptation becomes one of the major challenging issues to support the applications on different domains. In this paper, we propose a framework of domain-adaptable NLIs to integrate linguistic knowledge and world knowledge. Given a knowledge base of a target domain and the function definition of a target smart device, the corresponding NLI system is developed under the framework. In the experiments, we demonstrate a Chinese NLI system for a video on demand (VOD) service.

Keywords: Domain Adaptation, Knowledge-Driven Approach, Knowledge Graph, Natural Language Interface.

1 Introduction

Smart devices like smartphones, tablets, smart televisions, and intelligent appliances are widely popular nowadays. The most attractive feature of avant-garde smart devices is natural language interface (NLI), which makes many services such as controlling the devices and accessing the knowledge bases with natural languages feasible.

Constructing an NLI for a specific domain is a challenging task because the domain dependent linguistic knowledge and world knowledge have to be considered. For example, the entities used in an NLI to a traffic information system on a GPS client and in an NLI to a music on demand service on a cellphone are different. Besides, the terms to express the needs and the actions triggered by the terms to meet the needs for the two types of services are also different. Thus, how to adapt a system from one application domain to another is an important and urgent task.

This paper presents a framework for constructing domain-adaptable NLI systems. Section 2 examines the domain dependent issues in a framework of NLI to a knowledge base. Section 3 demonstrates a Chinese NLI system for a video on demand (VOD) service to show the feasibility of the methodology.

E. Métais, M. Roche, and M. Teisseire (Eds.): NLDB 2014, LNCS 8455, pp. 237–241, 2014.

2   Domain-Dependent Issues in NLI Systems

Assume the intent of a user in an NLI to a VOD service is to watch the movie "Romeo + Juliet", in which Leonardo DiCaprio acts the role Romeo. The user may submit a query like "I want to see Leonardo as Romeo". There are two entities in this query, i.e., the name of an actor (i.e., Leonardo) and the name of his role (i.e., Romeo) in a movie. The NLI system takes the following steps for this query: (1) to analyze the user's intent, (2) to check if there are any programs satisfying the need, (3) to identify which channel plays the program, and (4) to move to this channel.

To achieve the goal, the NL query is sent to a query processing module, where a query is segmented into a sequence of words, each word is labelled with a part-of-speech tag, and a dependency tree is generated for the query. The retrieval module tries to find the relevant information from the domain-specific knowledge base according to the analysis results. If the user intent is completely specified in the query, the corresponding action frame is triggered by the smart device. Otherwise, the human-device interaction is initiated to accumulate enough information.

In the above process, both linguistic knowledge and world knowledge are indispensable. Fundamental query processing needs the supports of linguistic knowledge from different levels. Entities, their properties, and relationships among entities are interesting targets to users. The domain-specific world knowledge, which is different from one domain to another, is created anytime and anywhere. The dynamic property makes linguistic analysis challenging.

To deal with this domain adaptation issue, knowledge in different domains can be formulated as a universal representation. With a uniform scheme, we can define common operations for all the domains. The concept of knowledge graph (KG) [1] is adopted. Entities are not only fundamental things in a KG, but also fundamental units in linguistic analysis. They serve as bridges between semantics and knowledge. Functions of smart devices are also domain dependent. Query term in a query implicitly triggers some function of a device. For example, the query term "see" requests an implicit action "move to a channel". Which query terms act as actions and how to map query terms into actions of devices should be defined for domain adaptation.

Furthermore, words unique to a specific domain will affect the performance of segmentation, part of speech tagging, and dependency parsing. Those domain-specific words are out-of-vocabulary (OOV) relative to a generic lexicon. Word forms and their POS are often adopted as features for a dependency parser. OOV words will affect the parsing results, and thus the subsequent retrieval performance. We adopt the knowledge-driven approach to collect domain-specific words from KG to decrease OOV problems. Besides, POS-based and word-rephrasing strategies proposed in our previous work [2] are explored to adapt a dependency parser.

3 An NLI System for a VOD Service

A Chinese NLI system for a VOD service is taken as an example. We will deal with the domain dependent issues discussed in the above in the following subsections.

3.1 Knowledge Base

The target domain data in the VOD service contain TV channels, news, movies, music, and other multimedia items. These items can be ordered and played on users' televisions. In addition, the metadata include the cast and crew of a movie, the singer of a song, the presenter of a television program, and so on. All the topics are the vertices in the KG, and the properties are the edges denoting the relations between topics. Fig. 1 shows the type of the topic *"The Dark Knight"* is *Movie*, and the associated topics including actors, directors, genres, subjects, released, and alias are the properties of this topic.

Fig. 1. The Movie "The Dark Knight" represented in a Knowledge Graph

3.2 Linguistic Analysis

In the experimental domain, the NLI should handle a lot of domain-specific named entities (NEs) like movie titles and actor names. Most of the NEs are created dynamically and are OOV terms in linguistic analysis. We integrate a named entity recognizer (NER) with Chinese word segmentation and part-of-speech tagging. The NE lexicon referred by the NER is derived from the KG for the specific domain. The NEs are not directly determined by the dictionary-based NER, but extracted from the decision of a learning-based segmenter.

Dependency parser determines the structure of a user query and labels the relations between the tokens. Fig. 2 shows the dependency tree of a Chinese query "我想看昨天的棒球經典賽" (I want to watch yesterday's World Baseball Classics). Dependency parser helps identify the core verb "watch" and the target entity "World Baseball Classics" in a query. We train a Chinese dependency parser with ZPar [3], an implementation of the transition-based dependency parser. OOVs have been segmented and tagged with the tag NR (proper noun) in the earlier steps. Referring to the KG, a sentence rephrasing approach [2] replaces an OOV word in a sentence with the thing of the same NE type in the training set, e.g., "baseball game", so that the knowledge of the word form can be used for parsing.

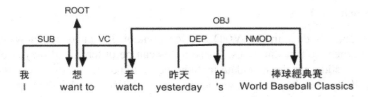

Fig. 2. A Dependency Tree of the Query "我想看昨天的棒球經典賽"

3.3   Action Frames

The action frames describe the functions of the target smart device. In the demonstration, the smart device is the controller of the VOD service attached to user's television. The functions of controllers include changing the TV channel, playing a movie or a song on demand, reporting the weather, and so on. One of the challenging issues is how users express their intents in this domain. Mining terms from users' query logs may be a solution, but logs are not always available before developing an NLI for a specific domain. Knowledge-driven approach is adopted to deal with this problem. Because entities, properties, and relationships in the KG are targets for retrieval, we consult external corpora to extract their collocated words, in particular, verbs. For example, "Romeo + Juliet", which is an entity in the KG, is a movie. The words such as "see" and "watch" are collocated with "movie" frequently, thus they are included in the intent list. Referring to the Chinese proposition bank (LDC2013T13), the verbs and their frame elements are collected, and mapped into actions for the smart device.

4   Conclusion

A knowledge-driven approach integrates linguistic knowledge and world knowledge for domain adaptation. Entities and their relationships in real world, which are usually domain-specific, serve as the complementary knowledge for linguistic analysis. Knowledge for the OOV words and the potential intents are extracted. A Chinese NLI application on a VOD service demonstrates the feasibility of the framework.

Acknowledgments. This research was partially supported by National Science Council, Taiwan under NSC101-2221-E-002-195-MY3 and 102-2221-E-002-103-MY3.

References

1. Singhal, A.: Introducing the Knowledge Graph: things, not strings. Official Google Blog (May 2012), `http://googleblog.blogspot.com/2012/05/introducing-knowledge-graph-things-not.html`
2. Huang, H.H., Chen, H.Y., Yu, C.S., Chen, H.H., Lee, P.C., Chen, C.H.: Sentence Rephrasing for Parsing Sentences with OOV Words. In: The 9th International Conference on Language Resources and Evaluation (LREC 2014), Reykjavik, Iceland (2014)
3. Zhang, Y., Clark, S.: Syntactic Processing Using the Generalized Perceptron and Beam Search. Computational Linguistics 37(1), 105–151 (2011)

Speeding Up Multilingual Grammar Development by Exploiting Linked Data to Generate Pre-terminal Rules

Sebastian Walter, Christina Unger, and Philipp Cimiano

Semantic Computing Group, CITEC, Bielefeld University, Bielefeld, Germany

Abstract. The development of grammars, e.g. for spoken dialog systems, is a time- and effort-intensive process. Especially the crafting of rules that list all relevant instances of a non-terminal, e.g. Greek cities or Automobile companies, possibly in multiple languages, is costly. In order to automatize and speed up the generation of multilingual terminal lists, we present a tool that uses linked data sources such as DBpedia in order to retrieve all entities that satisfy a relevant semantic restriction. We briefly describe the architecture of the system and explain how it can be used by means of an online web service.

1 Introduction

The development of grammars, e.g. for spoken dialog systems (SDS), is a costly process requiring large manual investments [2]. One aspect of SDS grammar development that is particularly costly is the process of developing pre-terminal rules, i.e. rules that expand a non-terminal into a number of named entities. For example, when developing a dialog system that provides access to the bus schedule connecting Greek cities, one needs a list of names of Greek cities. Similarly, when developing a dialog system that is able to perform conversions between different currencies, one needs a list of the relevant currencies that exist worldwide. In such cases we would like to create pre-terminal rules like the following ones:

```
1 GREEK_CITY = Athens | Thessaloniki | ...
2 CURRENCY   = Euro | Dollar | Yen | ...
```

The acquisition of lists of such entities is costly, and if the grammar needs to be developed for different languages, this problem is exacerbated.

On the other hand, a massive amount of structured and interlinked knowledge is currently emerging in the form of the Linked Open Data cloud[1].

In this paper we present an approach that supports the acquisition of lists of named entities in order to speed up the process of creating pre-terminal rules. The approach exploits DBpedia as the central hub of the Linked Open Data cloud. Given one or more examples, it retrieves the classes that those examples

---

[1] http://lod-cloud.net/

E. Métais, M. Roche, and M. Teisseire (Eds.): NLDB 2014, LNCS 8455, pp. 242–245, 2014.

share, as well as properties that they have in common. The user can select those classes and properties that are relevant, which are then used to retrieve all available entities and returns them in form of a grammar rule in ABNF format that can be integrated into the grammar project in question.

The paper is structured as follows. In Section 2 we provide some information about DBpedia. Section 3 then briefly describes our approach. We conclude in Section 4. The source code can be found at `https://github.com/swalter2/TerminalEnhancement`. To run the system a DBpedia SPARQL endpoint is needed, such as the official DBpedia SPARQL endpoint[2] with DBpedia 3.9. A link to a live demo of our system can be found in the GitHub repository.

2   Linked Open Data and DBpedia

The Linked Open Data cloud consists of a large amount of interlinked RDF[3] (Resource Description Framework) datasets, including knowledge bases such as DBpedia[4] and YAGO[5]. It has been growing steadily in recent years, now comprising more than 30 billion RDF triples[6].

The central hub of the Linked Open Data cloud is DBpedia [1], a cross-domain knowledge base that was extracted from Wikipedia infoboxes. The English version of DBpedia currently comprises around four million entities, most of them organized in a consistent ontology. This ontology includes labels in a range of languages. In addition, there are more than 100 localized versions of DBpedia available. YAGO is another knowledge base extracted from Wikipedia, which in contrast to DBpedia also includes information from WordNet[7] in order to categorize entities.

Such structured knowledge bases become more and more popular for various applications. However, to our knowledge, such structured knowledge bases have not been exploited to speed up spoken dialogue grammar development. In order to show the potential of the structured data available on the web for rapid grammar development, we present a tool that uses DBpedia in order to automatically generate terminal rules, given some semantic restrictions provided by a grammar developer.

3   Demonstration and System Description

This section briefly describes the tool that will be demonstrated at the conference. It uses one or more resources as input and returns a terminal grammar in ABNF format as output. Figure 1 provides an overview of the architecture.

---

[2] http://dbpedia.org/sparql
[3] http://www.w3.org/TR/rdf-primer/
[4] http://dbpedia.org
[5] http://www.mpi-inf.mpg.de/yago-naga/yago/
[6] http://lod-cloud.net/state/
[7] http://wordnet.princeton.edu/

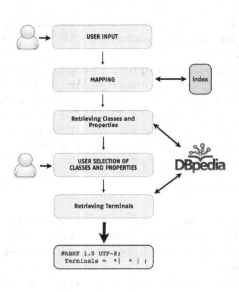

Fig. 1. System overview

First, the user has to enter the name of at least one (but optimally two or more) resources. In the following, we will use the input *Bruce Lee* and *Jackie Chan* as example. Currently this input is matched directly with DBpedia resource labels, but we will shortly add an index lookup that includes anchor texts from Wikipedia and thus allows also for indirect matches, e.g. using as input *Chan* or *Jacky Chan*. Given the input resources, the system retrieves all DBpedia and YAGO classes they have in common. In addition, all classes that belong to the DBpedia ontology namespace are sorted hierarchically, using the property `rdfs:subClassOf`. The most general class is presented at the top (in the case of *Bruce Lee* and *Jackie Chan* the class `Person`), the most special class is presented at the bottom (in this case the class `Actor`). In addition, the number of entities of the corresponding class are indicated in brackets after the class name. As the class names are not always very comprehensible, e.g. `Site108651247`, five example resources are provided for each class. The YAGO classes are not connected through `rdfs:subClassOf`, such that they are sorted by the number of elements they have. Furthermore, all properties for which the input resources share the object are returned. In the case of *Bruce Lee* and *Jackie Chan*, for instance, this would be the property `country` with object `China`.

The retrieval step works for an arbitrary number of (non-zero) input resources. This means that if only one resource is given, all types of that resource are retrieved, together with all property-object pairs of the triples it occurs in as subject.

Next, the user has different options to proceed:

1. choosing one or more relevant property-object pairs
2. choosing one or more classes
3. choosing one or more classes and also one or more relevant property-object pairs

If more than one class is selected, the user can furthermore choose if the semantic restriction should be created by connecting all classes using AND or OR semantics. The selected classes and properties are then used to retrieve all resources that match those constraints. Finally, the labels of those resources, possibly in different languages, are returned as the target terminals. Depending on the chosen constraint(s), a different number of resources (terminals) are returned. For our example of *Jackie Chan* and *Bruce Lee*, option one amounts to choosing `country` with object `China`, returning around 6,250 terminals. Option two would amount to selecting one or both of the classes `Actor` and `Person`. When connecting both classes by a logical AND, 2,669 terminals are returned; connecting them with a logical OR returns around 50,000 terminals. Option three returns 903 terminals when choosing both classes and the property-object pair, i.e. asking for all persons who are actors from China.

Besides the default language English, the user can currently choose one additional language, such as Spanish, German, Russian or Chinese. In case a terminal has only an English label, but none for the selected second language, only the English label is returned. In the demo we restricted the language choice for performance reasons, but in principle the system can return terminals in arbitrarily many different languages, as long as they are supported by DBpedia.

Finally, all returned terminals are displayed in the browser and for each language an ABNF grammar file is generated, which can be downloaded. We chose ABNF[8], as this is one of the most common SDS grammar formats, but the tool could be easily adapted to include other grammar formats as well, e.g. GRXML[8].

4   Conclusion

In this paper we presented a first approach of an easy-to-use tool that has the potential to significantly speed up the development of multilingual grammars by exploiting linked data for the generation of pre-terminal rules. As an evaluation of the tool, we plan to compare the amount of time needed by a grammar expert to create a terminal grammar by hand compared to the amount of time need when being supported by our tool. This will provide us a detailed picture of the efficiency of our approach.

Acknowledgment. This work has been funded by the European Union's Seventh Framework Programme (FP7-ICT-2011-SME-DCL) under grant agreement number 296170 (PortDial).

References

1. Bizer, C., Lehmann, J., Kobilarov, G., Auer, S., Becker, C., Cyganiak, R., Hellmann, S.: DBpedia – a crystallization point for the web of data. Web Semantics: Science, Services and Agents on the World Wide Web 7(3), 154–165 (2009)
2. Riccardi, G., Cimiano, P., Potamianos, A., Unger, C.: Up from limited dialog systems! In: NAACL-HLT Workshop on Future Directions and Needs in the Spoken Dialog Community: Tools and Data, pp. 1–2. Association for Computational Linguistics (2012)

---

[8] http://www.w3.org/TR/speech-grammar/

Senterritoire for Spatio-Temporal Representation of Opinions

Mohammad Amin Farvardin[2] and Eric Kergosien[1]

[1] LIRMM - CNRS, Univ. Montpellier 2,
161 rue Ada, 34095 Montpellier Cedex 5, France
eric.kergosien@lirmm.fr
[2] Cirad, UMR TETIS, 500 rue J.F. Breton, 34093 Montpellier Cedex 5, France
amin.farvardin@teledetection.fr

Abstract. In previous work, he method called OPILAND (OPinion mIning from LAND-use planning documents) has been proposed in order to semi-automatically mine opinions in specialized contexts. In this article, we present the associated SENTERRITOIRE viewer developed to dynamically represent, in time and space, opinions extracted by OPILAND.

Keywords: Land-use planning, Opinion, Spatio-temporal visualization.

1 Introduction

The characterization and understanding of perceptions of a territory by different users is complex, but needed for land-use planning and territorial public policy. In this paper, we focus on political and administrative territories (eg. local or regional territory referring to opinions, spatial and temporal features). In the context of land-use planning, even if information published on the web (blogs, forums, etc.) and in media express the feelings, the traditional opinion mining approaches, based on statistics or Natural Language Processing (NLP), fail to extract opinions due to the context specificity (small or medium-size and specialized corpus). To tackle this issue, we have proposed a new approach, called OPILAND (OPInion mining for LAND-use planning documents), to semi-automatically mine opinions in specific contexts. The approach is divided into three stages [4]: (1) semi-automatic extraction and disambiguation of named-entities (Spatial Feature (SF) and Organization) using a hybrid approach that includes a NLP sequence and supervised learning methods, (2) automatic extraction of specialized vocabulary sets that merge several French lexicons of opinions in order to compute a polarity score for a document, and (3) extraction of highly selective vocabularies of opinions related to the SF extracted during the first step and Temporal Features (TF : dates and periods) extracted using the NLP process defined by [5]. However, to the authors knowledge, there is no application permitting to represent opinions expressed by stakeholders on their territory. Therefore, we have developed a demonstrator called SENTERRITOIRE-VIEWER to visualize, in space and time, opinions extracted from documents.

E. Métais, M. Roche, and M. Teisseire (Eds.): NLDB 2014, LNCS 8455, pp. 246–249, 2014.
© Springer International Publishing Switzerland 2014

This work is part of a project called SENTERRITOIRE which aims at developing a decision-making environment for land-use planning documents.

This paper is structured as follows. In Section 2, an overview of opinion viewing applications is presented. In Section 3, the SENTERRITOIREVIEWER demonstrator is detailed. The paper ends with our conclusions and future work.

2   State-of-the-Art

Various works on opinion mining focus on topic classification. Some of them use visualization as a navigation tool [2]. Some approaches analyze datas relating to political views. Presidential Watch 2008[1] has launched a set of tools to see, hear and feel what citizens and supporters are saying on the Internet about the 2008 US presidential elections. The political blogosphere called BlogoPole[2], classifies over 2000 websites, related to the 2007 French presidential elections, by their political affiliation. In the field of news, [7] propose an approach based on the treemap algorithm to provide an overview of the aggregator Google News. The tool allows to compare the importance of various topics in newspapers. In social sciences, the White Spectrum tool [1] permits to analyze the debate generated by programs BBC2 in 2008. In the same vein, WeFeelFine [3] is an interactive website which identifies continuous expressions of emotion through blogs, and displays it dynamically in a scatter graph. OpinionSeek [8] also offers a visualization of opinions in a scatter graph automatically enriched from the continuous analysis of reviews posted by customers of a hotel chain. Closer to our work, LoveWillConquer is a project developed to celebrate Valentine's Day. It is a real-time, interactive, 3D stream of tweets per country that mentions "love" or "hate". A spatial filter is available but the visual representation in space is missing. [6] offers a 3 dimension browser that gives a spatial representation of opinions expressed on the European Community.

However, to the best of our knowledge, there is no tool representing opinions in the land-use planning context. Thereby, we have developed the demonstrator SENTERRITOIREVIEWER to navigate into a corpus according to three components: Opinion, Time and Space.

3   SenterritoireViewer to Dynamically Represent Opinions in Time and Space

The demonstrator integrates three components to dynamically visualize opinions in time and space:

- General visualization of opinions: general overview of opinions expressed in documents represented by a main pie chart is colored according to the percentage share of opinions (calculated automatically): red for negative, yellow for neutral and green for positive (cf. Opinion component in Figure 1). Five small

---

[1] http://politicosphere.net/map/
[2] http://blogopole.observatoire-presidentielle.fr/

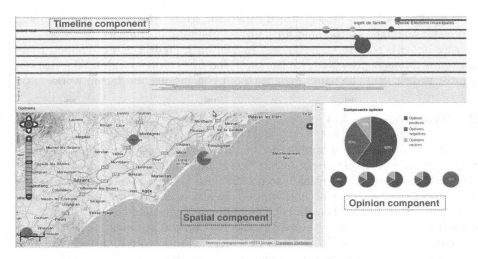

Fig. 1. The demonstrator SENTERRITOIREVIEWER

pie charts are added below the main graph to represent the evolution of the general opinion in time (the oldest opinion to the most recent opinion expressed in a given spatial and temporal context).

- Opinion in space: representation of opinions by geographical area described, using the cartographic web engine called Geoserver and web technologies (GeoExt, ExtJS and Open-Layers) for visualizing large amounts of data in vector or raster format. The module (cf. Spatial component in Figure 1) includes the following features: pan/tilt/zoom functions, select the desired scale, query objects with tooltip information about opinions, different base maps via the GoogleMaps API (Street, Satellite and Hybrid) and OpenStreetMap maps. For each pair Opinion - SF identified through the approach OPILAND, a pie chart is added to the map on the SF pointed. This pie chart follows the same characteristics as representing the overall opinion component. If several pairs are identified for the same ES, a single circular is displayed and includes all opinions. The size of each round is proportional to the number of opinions related to the ES pointed. For example in Figure 1, a majority of positive opinions are related to the Thau lagoon area.

- Opinion in time: representation of opinion by a timeline, using the TimelineJS API (cf. Spatial component in Figure 1). In the module, each document is represented by a line in time (from the first date to the last one identified in the content). For information on the document whose opinion is extracted, a system tooltip is implemented (visualization of the extracted sentence in which the opinion is expressed and URI of the document). Also, the temporal component incorporates the same principles as the spatial component: zoom with different scales of time (day, month; etc.), drag to scroll the timeline horizontally, colors, pie charts, grouping several opinions identified on the same date/period, etc.

The different components are dynamically linked. Specifically, the timing constraints specified in the timeline are taken into account by other components in their display. So, when we move in the timeline, the opinion and spatial components are automatically updated to display only the opinions identified in the relevant period. The Figure 1 highlights opinions extracted from documents (using OPILAND) in our corpus related to the Thau territory between March and June 2011. The experiments show that 83.82% of extracted SF are correctly identified and the vocabulary of opinions automatically built significantly improves the identification of document polarity (91.9%) [4]. The SENTERRITOIREVIEWER[3] thus provides information on the spatial environment to help experts of Syndicat Mixte du Bassin de Thau to diagnose their territory.

4   Conclusion and Future Works

We have proposed a demonstrator allowing to dynamically represent, in time and space, opinions extracted by the OPILAND approach [4]. Future work are allocated to evaluate the demonstrator by experts, especially experts of Syndicat Mixte du Bassin de Thau and local stakeholders. We are working on the definition of evaluation scenarios with geographers involved in the project SENTERRITOIRE.

Acknowledgments. The authors thank A. Abboute and G. Entringer (LIRMM, France), M. Roche, P. Maurel and M. Teisseire (UMR TETIS, France) for their involvement in the SENTERRITOIRE project. This work was partially funded by the labex NUMEV and the Maison des Sciences de l'Homme de Montpellier (MSH-M).

References

1. BBCTwo. Bbc news white spectrum (January 2008)
2. Fry, B.: Visualizing Data, 1st edn. (2008)
3. Kamvar, S., Harris, J.: We feel fine and searching the emotional web. In: Proceedings of the Fourth ACM International Conference on Web Search and Data Mining, WSDM 2011, pp. 117–126. ACM, New York (2011)
4. Kergosien, E., Laval, B., Roche, M., Teisseire, M.: Are opinions expressed in land-use planning documents? International Journal of Geographical Information Science (IJGIS) 28, 739–762 (2014); (Rank A, IF: 1.61 in 2012)
5. Lacayrelle, A.L.P., Gaio, M., Sallaberry, C.: La composante temps dans l'information gographique textuelle. Revue Document Numrique 10(2), 129–148 (2007)
6. Offenhuber, D.: Thought landscape (January 2005)
7. Ong, T., Chen, H., Sung, W., Zhu, B.: Newsmap: a knowledge map for online news. Decision Support Systems 39(4), 583–597 (2005)
8. Wu, Y., Wei, F., Liu, S., Au, N., Cui, W., Zhou, H., Qu, H.: Opinionseer: Interactive visualization of hotel customer feedback. IEEE Transactions on Visualization and Computer Graphics 16(6), 1109–1118 (2010)

---

[3] demonstration : `http://ekergosien.free.fr/file/projetSenterritoire.mp4`

Mining Twitter for Suicide Prevention

Amayas Abboute[1], Yasser Boudjeriou[1], Gilles Entringer[1], Jérôme Azé[1],
Sandra Bringay[1,2], and Pascal Poncelet[1]

[1] LIRMM UMR 5506, CNRS, University of Montpellier 2, Montpellier, France
{bringay,aze,poncelet}@lirmm.fr
[2] AMIS, University of Montpellier 3, Montpellier, France

Abstract. Automatically detect suicidal people in social networks is
a real social issue. In France, suicide attempt is an economic burden
with strong socio-economic consequences. In this paper, we describe a
complete process to automatically collect suspect tweets according to a
vocabulary of topics suicidal persons are used to talk. We automatically
capture tweets indicating suicidal risky behaviour based on simple classification methods. An interface for psychiatrists has been implemented to
enable them to consult suspect tweets and profiles associated with these
tweets. The method has been validated on real datasets. The early feedback of psychiatrists is encouraging and allow to consider a personalised
response according to the estimated level of risk.

Keywords: Classification, Suicide, Tweets.

1 Introduction et Motivations

According to the French website Sante.gouv.fr[1], nearly 10,500 people die each
year in France by suicide (3 times more than traffic accidents). Approximately
220,000 suicide attempts are supported by Emergency department. The economic burden of suicide is estimated at 5 billion euros for 2009 in France. Suicide is a major public health issue with strong socio-economic consequences. The
main objective of this study is to detect, as early as possible, people with suicidal
risky behaviour. To do this, we focus on recent information retrieval techniques
to identify relevant information in texts from the Twitter social network. These
messages are used to learn a predictive model of suicide risk.

Societal benefits associated with such a tool are numerous. The semi-automatic
detection model of suicidal profiles can be used by social web services providers.
For example, moderators can use such a model to prevent suicide attempts: by
communicating directly with the concerned person, by contacting relatives when
possible or by displaying targeted advertisements such as *SOS Amitié* (translation: SOS Friendship) advertisement which appears when users enter special
terms in google search. A detailed analysis of identified messages can also help
psychiatrists to identify emerging causal chains between socio-economic inequal-
ities and different suicidal practices.

---

[1] http://www.sante.gouv.fr/

E. Métais, M. Roche, and M. Teisseire (Eds.): NLDB 2014, LNCS 8455, pp. 250–253, 2014.

We will address three major challenges: 1) Building vocabulary to collect messages from Twitter social network and dealing with various topics related to suicide (e.g. depression, anorexia); 2) Mining messages which are extremely variable from the point of view inter and intra individual in order to propose a classification model to effectively trigger alerts and thus identify people with a high risky behavior; and 3) Presentation of suspect messages in a web interface for health professionals.

The challenges associated with this study are numerous because text analysis is difficult. Most of the NLP methods used in health domain have been applied to publications and hospitalization reports. Their transposition to tweets is far from trivial (limited to 140 characters texts with nonconforming grammatical structures, misspelling, abbreviations, slang). The originality of our solution is to be language-independent and to cover the entire knowledge extraction process: research, acquisition, storage, data mining, classification, visualisation of suspect profiles. To demonstrate the technological feasibility, a prototype is available online[2]. This type of approach can be generalized to widely varying textual data (e.g. blogs, forums, chat, email) and other areas (e.g. cyberbulling, natural disasters detection) for which it is important to identify behaviors known as abnormal based on lexicons.

2   Methodology

The method is divided into 4 steps:

1. **Vocabulary definition.** Thanks to [1][2], we identified 9 topics suicidal people generally talk about: *Sadness/psychological injury, Mental State, Depression, Fear, Loneliness, Description of the suicidal attempt*[3], *Insults, Cyberbullying, Anorexia*. We have defined manually a set of keywords related to these topics on specialized sites and obtained a vocabulary of 583 inputs. We have collected whole sentences that have been used in proven cases of suicide such as "I want to die" or "You would be better off without me".

2. **Suspects and proven messages.** We automatically collected a corpus of tweets containing the words of the defined vocabulary through the API Twitter (about 6000 suspicious messages). We also collected messages from accounts identified as those of persons having committed suicide (proved cases identified thanks to newspapers) (about 30 proven messages).

3. **Manual annotation of messages into two categories: risky tweets and non risky tweets.** Three computers scientists manually classify messages into the two categories (according to the information collected on proven cases). About 150 messages have then been used for the learning phase of the classification process.

4. **Automatic classification.** We automatically classified suspects tweets into risky and non risky tweets. Using WEKA[4], the performances of six classifiers

---

[2] http://info-demo.lirmm.fr:8080/suicide2/

[3] 70% of suicidal people describe concretely how they will realise the suicidal attempt.

[4] http://www.cs.waikato.ac.nz/ml/weka/

(JRIP, IBK, IB1, J48, Naive Bayes, SMO) were compared using a Leave One Out validation (LOO) and also with a 10-fold Cross-Validation (10-CV) and the results have been averaged under 10 iterations. We first apply no filter to attributes and then remove *Depression* that is very frequent in the "no risky tweet" and rare in the "risky tweets". Removing *Depression* has a significant impact for many classifiers but Naive Bayes remains the best one.

5. **Presentation of results via a web interface.** The objective is to enable psychiatrists to consult tweets indicating suicidal risky behavior, find the latest tweets of the profile and edit statistics.

3   Preliminary Results

Table 1 shows the results obtained with the different classifiers tested via WEKA. The best classifier is Naive Bayes both in LOO and 10-CV validation, with an accuracy of 63.15% in LOO and 63.27% in 10-CV. As the two classes have the same number of tweets, the accuracy's baseline is equal to 0% for LOO and to 47.33% for the 10-CV, which is significantly lower than 63.27% for Naive Bayes. Figure 1 shows the distribution of the different categories of the vocabulary in the 6,000 tweets collected and indicating suicidal risky behavior. The most represented categories are *Insults* and *Hurt* for risky tweets.

Table 1. Accuracy for different classification algorithms. Best results are in bold.

Dataset	baseline	JRip	IB1	IB3	J48	NB	RF	SMO
10 CV	47.33	55.37	60.16	55.24	57.14	**63.27**	61.03	60.56
LOO	0.00	47.04	60.53	48.68	44.74	**63.16**	59.14	60.72
Without Depression Attribute								
10 CV	47.33	61.23	62.38	61.03	58.65	**63.54**	62.34	60.66
LOO	0.00	61.78	60.00	61.18	57.24	63.16	**63.42**	59.34

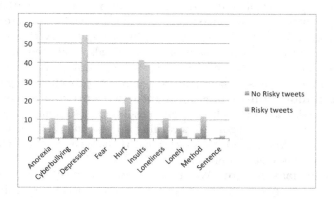

Fig. 1. Risky Tweets: Percentage by subcategories of the vocabulary

An online prototype was developed to start discussions with psychiatrists. The initial results were encouraging. They confirm that such a tool can be used to provide practical and efficient solutions for suicide prevention. They plan to base on care algorithms which could be personalised to take into account suicidal risk level (e.g. redirection to prevention websites, addressing to the nearest suicidal crisis unit, contacting relatives, etc.). These care algorithms must also integrate well-established risk factors (e.g. age, gender).

4   Conclusions and Prospects

In this article, we presented a complete process implementing automatic language processing and learning methods to identify Twitter messages indicating suicidal risky behavior. Initial feedbacks from psychiatrists are encouraging and allow them to define prevention methods customised by the level of risk. Prospects are numerous. We plan to collect more information about tweets to improve the learning phase of the automatic classification. First, we will extend the vocabularies with synonyms, antonyms to enlarge the scope of the suspects messages. Web statistic measures will be used to limit noise [3]. We will also use a general vocabulary of emotions to capture more special mental states [4]. Non-textual information such as the increase of the frequency of the tweets posting will also be taken into account. We will also improve the classification phase by using majority vote of multiple classifiers. We will obtain a list of tweets classified according to the level of risk. As our method is language independent, we will reproduce the study for French and Spanish to show its generality. For a medical point of view, we will conduct analyzes contrasting age, gender, location or any other information identified via user profiles Twitter.

Acknowledgement. We thank Ph. Courtet and S. Guillaume, Professors for their medical expertise.

References

1. Gunn, J., Lester, D.: Twitter postings and suicide: An analysis of the postings of a fatal suicide in the 24 hours prior to death. Suicidologi 17(3), 28–30 (2012)
2. Luyckx, K., Vaassen, F., Peersman, C., Daelemans, W.: Fine-grained emotion detection in suicide notes: A thresholding approach to multi-label classification. Biomed Inform. Insights 5(1), 61–69 (2012)
3. Roche, M., Garbasevschi, O.M.: WeMiT: Web-Mining for Translation. In: Conference on Prestigious Applications of Intelligent Systems, Montpellier, France, pp. 993–994 (August 2012)
4. Mohammad, S.M., Turney, P.D.: Emotions Evoked by Common Words and Phrases: Using Mechanical Turk to Create an Emotion Lexicon. In: Workshop on Computational Approaches to Analysis and Generation of Emotion in Text, pp. 26–34. ACL, Stroudsburg (2010)

The Semantic Measures Library: Assessing Semantic Similarity from Knowledge Representation Analysis

Sébastien Harispe[*], Sylvie Ranwez, Stefan Janaqi, and Jacky Montmain

LGI2P, École Nationale Supérieure des Mines d'Alès, Parc Scientifique G-Besse, 30035 Nîmes
`firstname.name@mines-ales.fr`

Abstract. Semantic similarity and relatedness are cornerstones of numerous treatments in which lexical units (e.g., terms, documents), concepts or instances have to be compared from texts or knowledge representation analysis. These semantic measures are central for NLP, information retrieval, sentiment analysis and approximate reasoning, to mention a few. In response to the lack of efficient and generic software solutions dedicated to knowledge-based semantic measures, i.e. those which rely on the analysis of semantic graphs and ontologies, this paper presents the Semantic Measures Library (SML), an extensive and efficient Java library dedicated to the computation and analysis of these measures. The SML can be used with a large diversity of knowledge representations, e.g., WordNet, SKOS thesaurus, RDF(S) and OWL ontologies. We also present the SML-Toolkit, a command-line program which gives (non-programmers) access to several functionalities of the SML, e.g. to compute semantic similarities. Website: `http://www.semantic-measures-library.org`

1 Semantic Measures Cornerstones of AI and Linguistics

The human ability to compare concrete and abstract objects (e.g., stimuli, concepts) is central in a wide array of cognitive processes. It has long been identified by cognitive sciences as a key element of learning and reasoning, among others [1][1]. In this context, assessing the similarity of object *representations* through algorithms has been of particular interest for the design of artificial intelligences. Therefore, several measures have been proposed to compare specific data (structures), e.g., number, strings, matrices, graphs. As an example, two images can be compared by analyzing their pixel matrices. However, in some cases, no simple *faithful* representations of objects can be used; this makes difficult to mimic the human ability to compare certain objects. Indeed, how to design a measure which will assess, as most people will agree, that the concepts `Peace` and `Dove` are more related than the concepts `Peace` and `Pigeon`? The measure must take into account the *meaning* of compared objects, i.e., it must consider that the concept `Dove` does not only refer to a bird, but also to a well-known (Christian) symbol of `Peace`, which explains that most people will define `Peace` and `Dove` as closely related.

---

[*] Corresponding author.
[1] Due to space constraint we will often refer to the extensive surveys [1, 2] for references.

E. Métais, M. Roche, and M. Teisseire (Eds.): NLDB 2014, LNCS 8455, pp. 254–257, 2014.

In this context, two strategies have so far been explored to compare objects by incorporating evidence regarding their meaning. The first strategy takes advantage of explicit Knowledge Representations (KRs) which define the meaning of terms, concepts and instances using a (more or less) formal semantics, e.g., semantic networks, logic-based ontologies. Such KRs will for example directly express that `Dove is-a Symbol-Of-Peace`. The second strategy proposes to capture the meaning of words or concepts by analyzing their implicit meaning in texts. This strategy extensively relies on the *distributional hypothesis* which states that words occurring in the same *contexts* tend to be semantically close [1, 2], e.g., *"Children released two white doves from Pope Francis' window as a peace gesture..."*, therefore, as the two words *dove* and *peace* will often co-occur, they will be regarded as closely related.

Thus, depending on the semantic proxy from which evidence of the meaning of compared objects is extracted, two broad types of *Semantic Measures* (SMs) are commonly distinguished: Knowledge-based SMs which rely on KR analysis [1] and Distributional SMs based on text analysis [2]. Due to the particular interest of numerous communities for SMs, a large diversity of measures have been proposed in the literature – please refer to the extensive collections of measures identified in [1, 2]. SMs are used in information retrieval, recommendation and question answering systems, to support knowledge inference, and to improve data mining treatments, to cite a few [1]. SMs are also of great importance for the NLP community in which they are used to resolve syntactic/semantic ambiguities [3], to generate ontologies from texts [4], to extract semantic relations between words/concepts [5], to summarize texts, to identify discourse structure, to detect plagiarism, to design systems, etc. [1].

Several tools have been proposed to compare lexical units using distributional SMs [1]. However, no generic tools have been proposed to compute knowledge-based SMs. Indeed, most of the developments in this area have been done for specific KRs; domain-specific implementations are numerous [1], e.g. WordNet [6], the Gene Ontology [7]. Thus, all the efforts made for specific KRs (e.g. algorithms implementations, unit tests, optimizations) cannot unfortunately be reused for other usage contexts. Thus, to overcome this limitation and to answer the need for efficient and generic software solutions dedicated to knowledge-based SMs, this paper introduces the Semantic Measures Library and associated toolkit.

2   The Semantic Measures Library and Toolkit

The primary goal of the Semantic Measures Library (SML) project is to give access to free, robust, generic and open source software solutions dedicated to SMs – at present, it mainly focuses on knowledge-based SMs. Two software solutions are proposed: the SML, a source code library, and the SML-Toolkit, a command-line program. They are distributed under the open source CeCILL license (GPL-compatible), and use the cross-platform Java programming language which is available for most operating systems (version 1.7). Downloads, documentation, tutorials, and community support for both the library and the toolkit are available on the dedicated website: http://www.semantic-measures-library.org.

The SML: A Source Code Library Dedicated to Semantic Measures. The SML is an extensive Java library dedicated to the analysis, development and computation of SMs. It provides numerous algorithms and measures implementations which can be used with a large diversity of KRs. The library is compatible with RDF(S), OWL, and SKOS, i.e. standardized Semantic Web languages commonly used for the definition of KRs. OBO ontologies, e.g., the Gene Ontology, and several domain-specific KRs are also supported, e.g., WordNet, Medical Subject Headings (MeSH), or SNOMED Clinical Terms [1]. In addition, specific KR loaders can easily be added to process other KRs using the library. Low-level access to the library enables developers to finely control the KR in order to apply specific treatments which are sometimes required for the computation of SMs (e.g. transitive reductions to remove taxonomic or annotation redundancies). This aspect is often essential to ensure the coherency of SMs' computation [1]. Interestingly, the graph data model used to process KRs guarantees the generic nature of the SML and therefore ensures that the implementations which rely on it (e.g., measures and algorithms) will benefit a broader audience and will not only target users of a specific KR.

The SML gives access to a large collection of measures, metrics and algorithms which enables hundreds of measure configurations to be expressed. Numerous state-of-the-art strategies proposed for the design of SMs are implemented [1]. These measures can be used to compute the semantic similarity, relatedness or distance between a pair of terms/concepts defined in a KR. Several measures can also be used to compare groups of terms/concepts, e.g., annotated documents or paragraphs.

The library implements the unifying abstract framework of SMs presented in [8] – it can be used to express a large number of SMs through parametric formulae, e.g., based on particular expression of the *ratio model* proposed by Tversky [1]. This framework is of particular interest for the large-scale analysis of SMs and for the selection of best-suited SMs in specific usage contexts [8].

The SML has been developed for large-scale computations and analyses of SMs. It supports fast parallel computations on multi-core processors. Moreover, despite its generic layer, the SML has been proved to outperform several domain specific libraries and tools dedicated to specific KRs, e.g., those dedicated to the Gene Ontology in the biomedical domain [7]. The library has already successfully been used in several (research) projects, e.g. to develop information retrieval and recommendation systems [1], to perform large-scale computations and analyses of SMs [7, 8].

The SML-Toolkit: A Command-Line Program. Numerous functionalities provided by the SML are also available within the SML-Toolkit, a command-line program which can be used (by non-programmers) to easily compute SMs on personal computers or computer clusters. The SML-Toolkit is highly tuneable and enables context-specific configurations w.r.t particular requirements of a usage context (KR, measure configuration, computational resources, etc.). Specific Command-Line Interfaces (CLIs) are developed in order to ease the use of the program, e.g., for WordNet, the Gene Ontology, the MeSH. Such domain-specific CLIs enable to hide the advanced capabilities of the library and can be associated with domain-specific documentations (using the terminology of the target community and thus enable end-users to focus on the important aspects of the domain use case). In addition, advanced and detailed configurations can also be specified using a generic XML configuration file. The performance of the toolkit have been shown to compete or outperform other KR-specific software solutions [7].

3 Conclusion

Semantic Measures (SMs) have been shown to be successful for a large variety of NLP tasks and are of particular interest for numerous communities (e.g., Linguistics, Semantic Web, Information Retrieval, and Biomedical Informatics). However, despite the interdisciplinary nature of this field of study, only domain-specific software solutions dedicated to knowledge-based SMs have been developed and adopted so far. This lack of generic software solutions dedicated to these SMs highly hampers their use, their study and their development. In response, and to federate the numerous communities which use and study SMs, this paper presents the Semantic Measures Library and associated toolkit: two efficient, generic, and open source software solutions dedicated to the computation, development and analysis of SMs. Downloads, documentation, supports and additional information are available at dedicated website http://www.semantic-measures-library.org.

References

1. Harispe, S., Ranwez, S., Janaqi, S., Montmain, J.: Semantic Measures for the Comparison of Units of Language, Concepts or Entities from Text and Knowledge Base Analysis. ArXiv. 1310.1285 (2013)
2. Mohammad, S., Hirst, G.: Distributional Measures of Semantic Distance: A Survey. ArXiv. 1203.1889 (2012)
3. Resnik, P.: Semantic Similarity in a Taxonomy: An Information-Based Measure and its Application to Problems of Ambiguity in Natural Language. J. Artif. Intell. Res. 11, 95–130 (1999)
4. Cimiano, P., Völker, J.: Text2Onto. In: Montoyo, A., Muñoz, R., Métais, E. (eds.) NLDB 2005. LNCS, vol. 3513, pp. 227–238. Springer, Heidelberg (2005)
5. Panchenko, A., Morozova, O.: A study of hybrid similarity measures for semantic relation extraction. In: Proceedings of the Workshop on Innovative Hybrid Approaches to the Processing of Textual Data, pp. 10–18 (2012)
6. Pedersen, T., Patwardhan, S., Michelizzi, J.: WordNet: Similarity: measuring the relatedness of concepts. In: HLT-NAACL, Demonstration Papers, Stroudsburg, PA, USA, pp. 38–41 (2004)
7. Harispe, S., Ranwez, S., Janaqi, S., Montmain, J.: The Semantic Measures Library and Toolkit: fast computation of semantic similarity and relatedness using biomedical ontologies. Bioinformatics 30, 740–742 (2014)
8. Harispe, S., Sánchez, D., Ranwez, S., Janaqi, S., Montmain, J.: A framework for unifying ontology-based semantic similarity measures: A study in the biomedical domain. Journal of Biomedical Informatics 48, 38–53 (2014) ISSN 1532-0464, http://dx.doi.org/10.1016/j.jbi.2013.11.006

Mobile Intelligent Virtual Agent
with Translation Functionality

Inguna Skadiņa, Inese Vīra, Jānis Teseļskis, and Raivis Skadiņš

Tilde, Vienibas gatve 75a, Riga, Latvia
{inguna.skadina,inese.vira,janis.teselskis,
raivis.skadins}@tilde.lv

Abstract. Virtual agent is a powerful means for human-computer interaction. In this demo paper, we describe a new scenario for mobile virtual agent that, in addition to general social intelligence, can perform translation tasks. We present the design and development of the intelligent virtual agent that translates phrases and sentences from English into French, Russian, and Spanish. Initial evaluation results show that the possibility to translate phrases and short utterances is useful and interesting for the user.

Keywords: virtual agent, multilingual information systems, mobile applications, automated translation.

1   Introduction

Nowadays, as mobile devices become ever more powerful, communication through virtual agents seems to be the most appropriate way to interact with a computer or mobile device. Several commercial applications, e.g., *Apple Siri* and *Google Now,* are the first important steps towards multimodal communication between a user and a computer.

Human-computer interaction has been actively researched for many decades. Different aspects, like dialog management, interactivity, reactive behavior, and others, are studied. Based on these studies, different applications are proposed – assistants, tutors, simple chatbots, etc. Although various virtual agents have been developed, translation tasks were not their primary focus. Some mobile phone applications provide multilingual translation services for a particular domain [1] without a multimodal component. Others have been developed as language teachers [2].

In this paper, we present an attractive 3D multimodal virtual agent Laura[1] for *Android* devices. Laura is a freely talking 3D head with natural mimics and synchronous lip movements, as well as emotions. The agent can answer questions and handle simple dialog based on AIML, *Artificial Intelligence Mark-up Language,* [3]. Besides answering a variety of questions, the agent can also translate words, phrases, and sentences from English into Spanish, French, and Russian.

---

[1] First version of Laura is available from https://play.google.com/store/apps/details?id=com.tilde.laura&hl=en

E. Métais, M. Roche, and M. Teisseire (Eds.): NLDB 2014, LNCS 8455, pp. 258–261, 2014.
© Springer International Publishing Switzerland 2014

2   Overview of the Virtual Agent

Our work is directed by the growing need for multilingual communication. Taking into account the wide use of mobile devices and user preference for speech input, we developed the virtual agent as a user friendly, socially intelligent, and freely talking 3D head. Although Laura can communicate about a wide range of themes, its novel function is a voice translation (Fig. 1.).

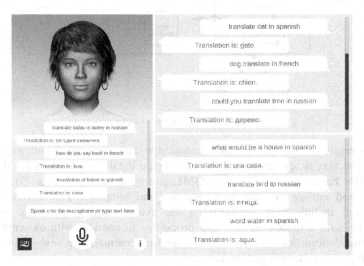

Fig. 1. Visual interface of the Laura application

The general architecture for the translation agent is shown in Fig. 2. It consists of five main constituents:

- Mobile application (implemented using the *Unity game engine* [4])
- Virtual agent web service (based on AIML language) – responsible for intelligent conversation between the agent and the user
- Automatic speech recognition (ASR) service – used for voice input
- Text translation service – used for translation of the user's queries
- Speech synthesis service (TTS) – used for speech output

Dialogue with the virtual agent can be managed through the voice command or typed in by using the keyboard. In the case of voice interaction, the user's request is recorded using the Unity engine microphone class to capture sound from the mobile device's built-in microphone. After being captured, audio is sent to the ASR service and received as a recognized text string. If the user prefers text input, text is sent directly to the Virtual agent web service that checks input patterns and gives appropriate answers accordingly.

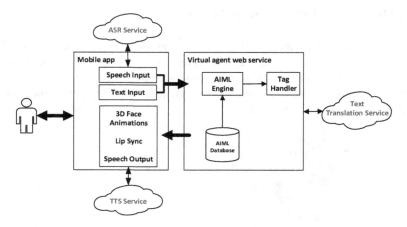

Fig. 2. Design of the system

Laura's replies are based on AIML knowledge that is extended with external web services. The text is sent to the AIML knowledge database to search for the proper answer. The basic structure of the AIML file is defined by tags [3], including beginning and ending tag, category, pattern, and template. The advanced AIML database can include tags for defining the avatar's emotions, tasks for AIML connectivity with a mobile device, and properties to connect with external services like text translation. For text translations, we implemented our new <translate> tag with two inputs for the text and the language.

Before implementation, a user study was performed to find out how people usually ask for translation. The most common phrases were "translate WORD to LANGUAGE" and "WORD translate in LANGUAGE", in addition to "Hi, could you, please, translate WORD in LANGUAGE".

An answer generated by the Virtual agent web service is sent to the TTS web service, and the user can hear the answer. To make the 3D model talk, we used language independent real time Lip Sync engine that dynamically creates mouth movements.

3   Evaluation

For evaluation, a questionnaire was used containing three groups of questions – information about the user's experience, questions related to the general features of the agent, and questions related to the translation task. 18 people between the ages of 20 and 50 (56% aged 20-30), with a gender breakdown that was 50:50, were involved. Most of the participants (61%) had experience using mobile devices for more than 3 years, while 28% used mobile devices for less than 1 year. Most of the participants (83%) thought that the agent is user friendly and 76% of participants were satisfied with the answers provided by the agent. Voice input was used by 83% of participants.

For the translation task, we analyzed two aspects – interaction and correctness of translation. Where it concerns the simplicity of interaction, 83% said "yes". To the question "Was the answer correct?", 11% responded "yes, always", 72% – "yes, mostly", 11% - "no, most of the time", and 6% said that the "answer was incorrect". However, for the question "Did Laura understood the question?", we got 67% negative answers (probably because of speech recognition quality), therefore we plan to analyse reasons of this contradictionary answer with help of additional evaluation.

Participants were asked to write down what they said to the agent. Most of participants used simple queries like "translate WORD to LANGUAGE" and "WORD translate in LANGUAGE". Most interesting queries were: "Dear Laura, could you, please, translate WORD in LANGUAGE" and "Do translation for WORD in Spanish language" or "Laura, please translate WORD in LANGUAGE".

In regards to what improvements users would like to see in the near future, most of the participants mentioned the addition of more languages. Some advised to make the application's primary function be translation, while others suggested it work more like a dictionary.

4   Conclusion

In this paper, we presented the intelligent virtual agent that translates phrases and sentences from English into French, Russian, and Spanish. Preliminary evaluation results show that the possibility to translate phrases and short utterances is useful and interesting for the user. Following user suggestions, we plan to add more languages for the translation task as well as make conversation with our agent multilingual.

Acknowlegments. The research leading to these results has received funding from the research project "Information and Communication Technology Competence Center" of EU Structural funds, contract nr. L-KC-11-0003 signed between ICT Competence Centre and Investment and Development Agency of Latvia, research No. 2.1 "Natural language processing in mobile devices".

References

1. Paul, M., Okuma, H., Yamamoto, H., Sumita, E., Matsuda, S., Shimizu, T., Nakamura, S.: Multilingual Mobile-Phone Translation Services for World Travelers. In: COLING 2008 22nd International Conference on Computational Linguistics: Demonstration Papers, pp. 165–168. COLING Demos (2008)
2. Hazel, M., Mervyn, A.J.: Scenario-Based Spoken Interaction with Virtual Agents. In: Computer Assisted Language Learning, vol. 18, pp. 171–191. Routledge, part of the Taylor & Francis Group (2005)
3. Wallace, R.S.: The Anatomy of A.L.I.C.E. In: Epstein, R., Roberts, G., Beber, G. (eds.) Parsing the Turing Test: Philosophical and Methodological Issues in the Quest for the Thinking Computer, pp. 181–210. Springer (2008)
4. Unity game engine, https://unity3d.com/unity

A Tool for Theme Identification in RDF Graphs*

Hanane Ouksili, Zoubida Kedad, and Stéphane Lopes

PRiSM, Univ. Versailles St Quentin, UMR CNRS 8144, Versailles France
`firstname.lastname@prism.uvsq.fr`

Abstract. An increasing number of RDF datasets is published on the Web. A user willing to use these datasets will first have to explore them in order to determine which information is relevant for his own needs. To facilitate this exploration, we present a system which provides a thematic view of a given RDF dataset, making it easier to target the relevant resources and properties. Our system combines a density-based graph clustering algorithm with semantic clustering criteria in order to identify clusters, each one corresponding to a theme. In this paper, we will give an overview of our approach for theme identification and we will present our system along with a scenario illustrating its main features.

Keywords: Theme idendification, RDF data, Clustering.

1 Introduction

An increasing number of RDF datasets is published on the Web, making a huge amount of data available for users and applications. In this context, a key issue for the users is to locate the relevant information for their specific needs. A typical way of exploring RDF datasets is the following: the users first select a URI, called a seed of interest, which they are willing to use as a starting point for their queries; then they explore all the URIs reachable from this seed by submitting queries to obtain information about the existing properties.

To facilitate this interaction, we present a system which provides a thematic view of a given RDF dataset in order to guide the exploration process. We argue that once the data is presented as a set of themes, it is easier to target the relevant resources and properties. Our system combines a density-based graph clustering algorithm with semantic clustering criteria in order to identify clusters, each one corresponding to a theme.

In the following, section 2 presents the general principle of theme identification. Our prototype and some scenarios are illustrated in section 3. Finally, section 4 concludes the paper.

2 General Principle of Theme Identification

Given an RDF dataset, our goal is to identify a set of themes and to extract the labels or tags which best capture their semantics. Providing this thematic view raises sevral questions:

---

* This work was supported by Electricity of France (EDF R&D).

E. Métais, M. Roche, and M. Teisseire (Eds.): NLDB 2014, LNCS 8455, pp. 262–265, 2014.

- Which information could be used to define a theme?
- As different users may not have the same perception of the data, how to capture their preferences and use them for building the themes?
- Finally, once the themes have been identified, how to label them so as to make their semantic as clear as possible to the user?

Our approach relies on the idea that a theme corresponds to a highly connected area on the RDF graph. The more a set of resources is connected, the more likely it is that they belong to the same theme or are related to the same topic. We will therefore use the structure of the RDF graph itself in order to build the themes. We apply a graph clustering algorithm which identifies these highly connected areas and their neighbourhood in order to form clusters, each one corresponding to a theme.

The structure of the graph alone is not sufficient to provide meaningful themes. Indeed, different users may have distinct perceptions of what a theme is. If we consider a dataset providing information about universities and scientists, one possible view is that themes correspond to research areas such as Mathematics or Physics, another one is that themes corresponds to research teams located in the same geographical area. These preferences will be used for identifying the themes, in addition to the structure of the graph.

Users preferences are captured by specifying a set of resources which should be assigned to the same cluster (for example, resources having the same value for a given property, or linked by a given property). Each preference will be mapped into one or several transformations applied to the graph. For example, if the user express that two resources related by the *sameAs* property should be assigned to the same cluster, the transformation will consist in merging the corresponding nodes in the graph.

3 Prototype and Scenarios

In this section, we illustrate the way our system discovers themes from an RDF dataset with four scenarios. We use a subgraph extracted from DBPedia (see figure 1). This data set contains resources describing scientists working in different domains with their organizations and their countries.

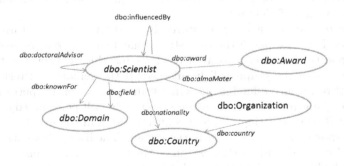

Fig. 1. Description of the RDF Dataset

Our tool enables theme identification by combining a density-based clustering algorithm with the specification of semantic criteria capturing user preferences. We have implemented the clustering algorithm proposed by [1], based on the notion of *k-core* [2]. The system requires two types of parameters: (i) clustering parameters, used to specify thresholds for assigning a node to a cluster, and (ii) semantic parameters, used to capture users' preferences.

The following interactions illustrate the way our tool is used for theme identification:

Scenario 1. The user wants to explore the RDF graph, but he has no knowledge about its content. The identification is done without taking into account any preference. The result is a set of clusters, each one corresponding to a given theme. Each cluster corresponds to a research domain but some resources related to the same domain have been allocated to different clusters.

Scenario 2. The user wants to perform a new identification task, but would like scientists from the same domain to be assigned to the same cluster. He will specify semantic parameters. In our example, the research domain is represented by the *field* property, and the user will indicate that two resources having the same value for this property should be assigned to the same cluster. The user can repeat the clustering process either on the initial graph by adding new semantic parameters, or on a cluster obtained in previous iterations in order to get further details.

Scenario 3. According to the preferences specified in the previous scenario, scientists of the same domain will be assigned to the same cluster. It may happen that this property is not defined for some of the scientists in the dataset, and the user would therefore like to use another semantic criteria. For example, he could state that scientists related by the *doctoralAdvisor* property should be assigned to the same cluster.

Scenario 4. In scenario 2, the user was interested in grouping scientists from the same research field. But another user could consider that the location should be considered instead for grouping resources. In such case, the semantic critera could be the values of the property *country*.

In our system, preferences are captured by modifying the structure of the internal graph. For example, if the user states that two resources related through a given property should be allocated to the same cluster, then the two corresponding nodes in the graph will be merged. If the semantic criteria specifies that a set R of resources having the same value for a given property should be allocated to the same cluster, then the system adds all possible edges between resources of R, thus creating a highly connected area in the graph.

Figure 2 shows the user interface of the system. The list of clusters is displayed on the left side and the initial RDF graph on the right side. The cluster selected in the list can be highlighted on the graph (a light nodes) or opened as a new RDF graph. In figure 2, the selected cluster corresponds to the field of *Mathematics*.

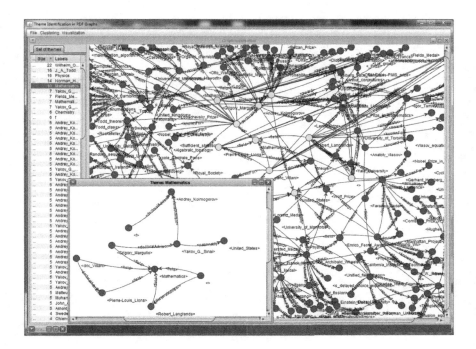

Fig. 2. Visualization of the themes

4   Conclusions

We have described the general principle underlying our prototype for theme identification in an RDF graph, which mainly comprises three stages: (1) pre-processing and capturing user preferences, (2) density-based clustering to form the clusters and (3) extraction of labels to describe the semantic of the cluster. Our approach differs from existing ones such as [3] in that it combines structural and semantic criteria for graph clustering. We are currently extending the system by improving label identification for a given cluster.

Acknowledgments. We would like to thank Sylvaine Nugier, Geoffrey Aldebert and Mokrane Bouzeghoub for their frequent feedback and advice.

References

1. Bader, G., Hogue, C.: An automated method for finding molecular complexes in large protein interaction networks. BMC Bioinformatics 27, 1–27 (2003)
2. Bollobás, B.: Graph theory and combinatorics. In: Proceeding of the Cambridge Combinatorial Conference in Honor of Paul Erdos, vol. 43 (1989)
3. Castano, S., Ferrara, A., Montanelli, S.: Thematic clustering and exploration of linked data. In: Ceri, S., Brambilla, M. (eds.) Search Computing III. LNCS, vol. 7538, pp. 157–175. Springer, Heidelberg (2012)

Author Index